VIDEOLOGY
Theory and Techniques

VIDEOLOGY
Theory and Techniques

by

H.C. Tien, M. D.

VANTAGE PRESS
New York Washington Atlanta Hollywood

Photography by
Timothy J. Telechowski

Published by Vantage Press, Inc.
516 West 34th Street, New York, New York 10001

Manufactured in the United States of America
Standard Book Number 533-03138-9

CONTENTS

INTRODUCTION

1. HISTORY AND EXAMINATION 1
2. TV-MONITORED MEDICAL FORMS 9
   Initial History and Examination (Forms 1 & 2)
3. TV-MONITORED MEDICAL FORMS 16
   Past Problems, Previous Treatment, and Personal History (Forms 3, 4, & 5)
4. TV-MONITORED MEDICAL FORMS 24
   Family History, Systemic Review, and Trend and Content Questionnaire (Forms 6, 7, & 8)
5. VIDEO-MONITORED MEDICAL HISTORY 31
   Four Cases
6. PSYCHOSYNTHESIS 46
   Psychoanalysis with Telefusion
7. TELEFUSION COUCH THERAPY 53
   Primary Telefusion, Ordinary Telefusion, Booster Telefusion
8. FROM THE COUCH TO THE COFFEE SHOP 59
9. FIRST TV-LINKED GROUP THERAPY 63
   Moreno, Freud, and TV-Linked Family Psychiatry
10. SECOND TV-LINKED GROUP THERAPY 73
    Television and Family Medicine
11. THIRD TV-LINKED GROUP THERAPY 84
    First TV Group Booster
12. TV-LINKED CYBERNETIC SYSTEM OF PSYCHOTHERAPY 92
13. A TV-CYBERNETIC HOLOGRAM MODEL 98
14. THE PERF MODEL OF VIDEOSYNTHESIS 104
15. TV-LINKED FAMILY PRACTICE OF MEDICINE 113
16. TV-LINKED GROUP THERAPY REVISITED 117
    Family Medicine and Videologists
17. PSYCHOSYNTHESIS, VIDEOTHERAPY, AND PSYCHIATRIC MANPOWER 127
18. VIDEOLOGY AND MEDICAL ECONOMICS 132
    Michigan Relative Value Studies
19. VIDEOTECHNIQUES AND VIDEOSYSTEMS 143
    CCTV Cameras, Video Cartridge Tape Recorders, and CCTV Monitors

20. VIDEOLOGY AND THE FUTURE PRACTICE OF MEDICINE 156
A PSYCHOSYNTHETIC GLOSSARY 160
APPENDICES 177
Initial Contact Form. TV-Monitored Evaluation Forms, 1 and 2; 3, 4 and 5; 6, 7 and 8 (Part 1 and Part 2); Neurological Examination (8a); Physical Examination (8b); Modified Medical Schedule; Depression Inventory Measure (D.I.M.); Psychiatric Prognostic Scale (P.P.S.); PQ—The Problem Questionnaire; Notes TV-Treatment Programs; Notes on Telefusion Couch Feedback; Progress Report.

BIBLIOGRAPHY 201
INDEX 211

INTRODUCTION

The psychiatrist, like the general practitioner, is moving gradually from solo practice to family practice in the United States, that is, he is becoming a family psychiatrist, accepting the mental health responsibilities of other members of a family in his professional care. However, he needs modern techniques to shoulder the added duties.

Every family psychiatrist must concern himself with the problems of time demands placed upon him, costs and effectiveness of his therapeutic programs, regardless of his orientation or school of thought. The solutions and answers to some of the questions raised may well lie with the use of television equipment and modern videotechniques.

This manual will demonstrate the use of videotapes and videotechniques as actually practiced in the everyday clinical situations of individual, marital, group, or special therapies. Physicians with experience in videotherapy are invited to consider in using video-equipment in the medical history-taking, interviews, couch sessions, marital or group sessions. Complex techniques can be shown on video, for example, the use of cross-feedback of couch sessions between the wife and husband in therapy who normally cannot communicate peacefully at home, but they would be able to listen to each other more carefully and tolerantly on videotape.

In brief, this manual will provide an opportunity for all interested to learn and to update information on the modern techniques of family psychiatry: videology.

Videology is the science and art of medical television. Videology is analogous to radiology, in that here we employ the use of camera and videotape for the examination and diagnosis and therapy in the treatment and prevention of disease and safeguarding of health, as we have in radiology in the employment of the X-ray machine and film in internal medicine and surgery.

Videotherapy equipment to the family psychiatrist may become as inseparable as the X-ray machine is to the radiologist and the microscope to the pathologist. Modern techniques make family psychiatry dynamic and effective.

A psychiatrist cannot simply sit behind the analytic couch, taking care of only the urban rich or the suburban well-to-do; or to service only the people of industrialized nations but not the developing countries. A psychiatrist should treat the working families as well as the middle-class.

With the limitation of time, how can a psychiatrist do it?

The psychiatrist must consider becoming a family psychiatrist with the use of modern technology—videology. In fact, videology has developed so extensively in the practice of family psychiatry that individual couch therapy can be applied on a mass scale, offering marital therapy, family therapy, and individual psychotherapy with special emphasis for child-rearing and guidance.

Family psychiatry recognizes that human beings need other human beings. It begins very early with child-rearing. The children when very young do not let the parents out of their sight, and the wise parents usually make the proper provisions for them. It is true for the family psychiatrist, who must talk to the adults as well as their children. We have to have time to give the quality of service. How are we to maintain the quality of communication and availability of ourselves?

The questions and the answers would be the central theme of this manual: how to use videotapes and videotechniques in the practice of family psychiatry and family medicine.

Finally, with the use of videotherapy, based on a reasonable theory of information transfer and communication, the solo psychiatrist becomes a family psychiatrist as a primary physician who has time to practice this art and science of medicine more fully. And the family physician may use the same videotechniques to improve his efficiency and effectiveness by educating his patients and families on the reality of medical service with videotherapy to enhance communication and cooperation for all interested in improving the health delivery system for the people.

H. C. Tien, M.D., Director
Michigan Institute of Psychosynthesis

Lansing, Michigan
May 1, 1977

VIDEOLOGY
Theory and Techniques

1. HISTORY AND EXAMINATION

This chapter will demonstrate how television can be used as a monitor for obtaining medical history in examining patients. The techniques are simply a direct application of the POPF Model.

The POPF Model (acronym for Professional contact-On television-Professional contact-Feedback) is derived from the PERF Model for TV-linked psychotherapy. The POPF Model can be applied not only to psychotherapeutic sessions but also to any learning situation, whether between doctor-patient, counselor-client, teacher-student, family lawyer-family member, when personal contact can be meaningfully amplified by sandwiching TV tapes between personal contacts with proper professional feedback.

In other words, the psychiatrist can see the patient in person at the beginning of a TV-linked therapeutic session to elicit the chief complaint or problems that day; then the patient is left with the same psychiatrist on TV to amplify the therapeutic information. At the completion of a half-hour TV treatment program, the psychiatrist returns for further personal contact and relevant transaction regarding future appointments, medication, testing, etc., so that the loop is completed with feedback. The POPF Model can also be applied to couch therapy, family therapy, interview psychotherapy, hypnotherapy, etc. The general implication of the POPF Model is that one physician can become two physicians, or more, so that it may be an answer to the medical manpower shortage.

By using the POPF Model, we are not here eliminating the physician but in fact intensifying the personal contact. The physician introduces himself to the patient on television and the television personality of the physician begins the TV-monitored medical history. The TV-monitored history is then filled in by the patient, who answers item by item as the physician or psychiatrist obtains the detailed complete medical background of the patient before instituting therapy.

The TV-monitored medical history is, of course, a great timesaver. Moreover, the physician is always calm, present, and able to go through all the necessary details to obtain a good analysis of the patient's past in the service of his present needs. This TV-monitored approach enables the physician to take several medical histories within the same hour, at the same time being thorough and complete. The TV system is adaptable to all physicians and each medical history would be done as the physician is when he is at his best, when he is prepared, when he is not fatigued, and is willing and consistent in his medical procedures. Since the physician always performs his best, which is the one that has been recorded for future TV-monitored patient histories, then we would have a consistent product and result as needed.

This article shows how completely the television monitor can duplicate the personality of the physician, to make his medical knowledge and service available to all who need medical care—as a right and not as a privilege.

The TV-monitored evaluation forms on these pages are used to obtain the chief complaint and evaluation of current problem, past problems, previous treatment, personal history, family history, systemic review, and trend and content of the patient's thought.

The following dialogue is excerpted from three television tapes used to obtain medical histories. One half-hour videotape is used for Forms 1 and 2; another for Forms 3, 4, and 5, and the third for Forms 6, 7, and 8. A different patient is being interviewed on each tape. Every question is read aloud, with time allowed for the patient's responses, which are written down on the printed forms.

INITIAL HISTORY AND EXAMINATION

DOCTOR: What is the chief complaint? What would you say, just a few words?
PATIENT: Something's wrong with me.
DOCTOR: Okay, put down "something's wrong with me." If you have any symptoms or signs, like crying spells, write that down. Now, you see under "B," how long have you been concerned about this, approximately?
PATIENT: I thought about going to a psychiatrist a long time ago. Maybe six years ago. About two weeks ago I become really positive about coming to see a psychiatrist because I don't think I can make decisions myself, can't grow up. It's hard to explain.
DOCTOR: In "B" we ask, "Perhaps you can't be sure of the problem, but what really bothers you?" What we mean by that is, how it affects you, bodily, physically, emotionally, or family-wise. How does it disturb you or make you uneasy?
PATIENT: It just pops out of nowhere. Things can be fine, then I start feeling fear.
DOCTOR: How does it bother you?
PATIENT: Makes me nervous, can't eat, sick to my stomach.
DOCTOR: Next, is it connected with your family in any way?
PATIENT: What if you think it is partly?
DOCTOR: Just mark Yes and write below, Partly. Who is this family member involved? Write down the person's name and how.
PATIENT: Partly. My sister, what happened to her with getting a divorce, getting sick, that upsets me.
DOCTOR: Now, J is a very important question: "Have you had a similar situation before?" Yes, No, Don't know. How did you answer?
PATIENT: Yes.
DOCTOR: Then put down a short description, about the time or the situation. Every case is different. Just whatever applies to you.
PATIENT: It's vague.
DOCTOR: Do you have anything else you are looking for? Do you feel we have covered everything? There is a blank page there and you can write down anything you feel may be of importance to me we have not covered. Have we missed anything? If we have, put it down. Well, we've gone through quite a bit of territory. Write down if you think of anything else.
PATIENT: I would like to talk about drugs and how they affected my thinking and my thoughts about myself.

PAST PROBLEMS; PREVIOUS TREATMENT; PERSONAL HISTORY

DOCTOR: Have you had the same or similar problems before, like headaches, or whatever troubles you may have?

PATIENT: I've been having headaches, but now they are getting worse.
DOCTOR: Give me the approximate dates, when you first started to have the symptoms. Can you describe the symptom, too, a little bit? Was it very severe? Did you have to quit work? Were there associated symptoms? Question number 2, have you been to any other psychiatrist, a family psychiatrist or psychiatrist of any kind? Yes or No?
PATIENT: You mean for myself?
DOCTOR: Or for anybody.
PATIENT: I talked to one concerning my daughter.
DOCTOR: Would you put that down, and your daughter's name, so I have some idea. What was the problem about? That may give me a clue what you worry about. The daughter worried the mother.
PATIENT: Right now she isn't home. I'm taking care of her kids plus my own, and she can't be found. That worries me.
DOCTOR: Yes, why didn't you continue to work with these doctors? We know that to solve problems requires time and effort and continuity. Why didn't you continue to go?
PATIENT: I began to seek for a deeper depth, more in the spiritual vein.
DOCTOR: Continuing on with Previous Treatment, have you ever had nerve pills or tranquilizers? There is a chart to list them. Do you know the name? Just put down the color and shape if you don't.
PATIENT: What about ones from the drugstore?
DOCTOR: Yes. Self-medication is of interest to me.
PATIENT: Sometimes I take a lot of aspirin.
DOCTOR: Now we will go in a little deeper with Personal History. The first thing to answer is about conception. That means how you were conceived, did your mother carry you normally, Yes or No? If it was not normal, what was the trouble?
PATIENT: I was premature. A seven-month baby.
DOCTOR: Now your philosophy. How do you look upon life, what is life all about?
PATIENT: Life is thinking of others. Not from a selfish point of view.
DOCTOR: Then we go into outlook on life. Do you think life is a great future, no future, absurd, interesting? Are you a pessimist, optimist, realist?

FAMILY HISTORY; SYSTEMIC REVIEW; TREND AND CONTENT

DOCTOR: How about your mother? Living? What age? List her education, occupation—that is, what she considers herself to be, whether it is housewife, teacher, or whatever. If she was married before, you can note that. How is her adjustment, good, fair, or poor?
PATIENT: Good.
DOCTOR: Is there any comment you want to make, anything about her personality relative to you, or your relationship? You don't have a stepmother or adopting mother?

PATIENT: No.
PATIENT: What if my brother got a divorce, and then remarried. Should I just list him as married?
DOCTOR: Yes, but also put down the divorce, or make a comment. We can explore that later on, how you reacted to it.
DOCTOR: Continuing more on Systemic Review, have you ever had any nerve disorder?
PATIENT: I don't know. I thought maybe that was what was giving me headaches.
DOCTOR: Put down how many years you have had these headaches. Write down any head injury, head trauma, if you were ever knocked out, or had convulsions or spells, loss of consciousness, passed out, fainting.
PATIENT: Not that I recall.
DOCTOR: Number 19: "Have any of your parents, brothers, sisters or near relatives ever had diabetes, mental illness or tuberculosis?"
PATIENT: My father had three nervous breakdowns.
DOCTOR: Now to the next form. Trend and Content Questionnaire. Number 1, what do you worry about most of the time? Everybody worries about something, so what do you worry about? Put down whatever comes to your mind; reject nothing.
PATIENT: Money and the military draft.
DOCTOR: Does anything depress you, make you blue? Do you get unhappy, brood about it?
PATIENT: Yes, but nothing in particular.
DOCTOR: Next, I am going to give you three wishes, any wishes you want. What three most important wishes would you have in your whole life? Any wish you can think of. Here you can just absolutely let your imagination take over.
PATIENT: 1. Health for my wife and I. 2. A comfortable life. 3. Wisdom.

DISCUSSION

The reflex reaction of those who are not familiar with the new concept of videology is to think in terms that the "cycloptic TV monitor" will take over personal contact. This is an important point. It is being eliminated in the sense of 1984, of George Orwell. In fact, on the contrary, this is just the opposite. We are using TV technology with increased information and on an informal level, rather than through the state or any sort of public television. Someone has called it "personal" television, which answers the needs of the families and at the same time frees the professional from the routinized yet important data collection from members of the family, whether in medicine, psychiatry, law or counseling.

When time has been created for the professional from routinized work, then he will be able to bring information to the families directly for their immediate problems, such as emergencies, crisis intervention, detailed explanation, personal rapport, and group participation, which in turn can be TV-taped for other members for mass communication and improved medical information at the lower cost in the long run for both treatment and prevention.

The idea of personal contact sometimes becomes a ritual, or a fetish, but the idea that unless the professional is "eyeball-to-eyeball" with the person then no benefit has been derived is patently away from the truth. A classical example is obvious. The old-fashioned doctor used to hold hands with the patient while the patient was dying of pneumonia, holding hands through a crisis but the patient died. Today's physician often sees a pneumonia patient for a much shorter time in a hospital, while the patient is given a series of penicillin injections, without the time-consuming eyeball-to-eyeball or handholding rituals in order to gain recovery. True, the amount of time in certain contacts is significantly reduced, but the professional time saved is actually available in many other directions. As a consequence, the quality of personal contact is improved and made appropriate for serving the real welfare of the people and community.

Furthermore, we must look upon the larger picture. If the physician, psychiatrist, or lawyer does not have time to see families at all, how much more impersonal the world would be.

With the whole approach we have made possible, both economically and time-wise, the evaluation of multiple family members in the same session. For example, if a couple comes in and the wife says the husband needs evaluation, and the husband says, "Oh, no, she is the sick one," we can settle the issue right away by modern technology. Both can be evaluated simultaneously if impartially by TV-monitored approach, while the professional person would still have time to do more objective analysis, thinking and assessment when the data are in.

We might also add a note here that TV-monitored approach is helpful to those people who cannot read or write, or who have difficulty in doing so. They can listen to the questions on tape, respond verbally, and their answers on tape can be transcribed by a secretary. A TV programmer can be present to assist the family that needs help because of educational or physical limitations.

SUMMARY

The science of videology or medical television technology came out of practical and theoretical attempts in bringing electrical engineering into the practice of psychiatry, where the medical manpower shortage has been most acute. Happily, the results are applicable to the practice of medicine in general.

The author has reported a ten-year (1961–1970) theoretical research between electrical engineering and modern eclectic psychiatry and a three-year (1968–1971) practical experience of building a TV-linked system of family psychiatry, fusing 260 families under his care into a mental health community. The method is entirely TV-taped for objective study and analysis, and the results are adaptable to the private practice of a solo psychiatrist or to a community mental health center.

The TV-linked system of family psychiatry has developed rapidly, giving rise to different types of therapeutic sessions, such as TV-taped couch sessions, TV-linked sessions, TV booster sessions, and telefusion couch therapy. Telefusion couch therapy is a most exciting development, in that the patient can undergo analysis and synthesis with television feedback, enabling one psychiatrist to analyze and synthesize two or more patients simultaneously in the same therapeutic hour. As described in this article, the TV-linked techniques are also applied to history-taking, psychological testing, and all therapeutic procedures.

The author makes extensive use of closed-circuit television for the private practice of family psychiatry, with eight channels and sixteen consultation rooms. This is a TV-system approach for the practice of modern eclectic psychiatry. A TV-linked system can actually service eight families per therapeutic hour, by one solo psychiatrist, by the use of paramedical personnel and television technology in a system approach. This new approach can improve the therapeutic efficiency of a psychiatrist by 316% and economy by 25% in a ten-year follow-up study (1961—1970). One of the most significant advances in TV-linked family psychiatry is that the psychiatrist can keep his office open during his short absences for professional or personal reasons, because the therapeutic procedures can be carried on through the help of para-psychiatric personnel together with the psychiatrist's personality on the TV monitor. In fact, even couch analysis can be continued by the modern techniques.

As a result of extensive application of television technology and TV tapes, the author can treat many times the number of patients per week as compared with eclectic approach alone. Cogently, the author has found that one psychiatrist literally becomes two psychiatrists in the TV-linked family practice of medicine.

REFERENCES

1. H. C. Tien, "TV-Linked Family Practice of Medicine." *World Journal of Psychosynthesis* 3:4–5.
2. H. C. Tien, "Television and Psychiatric Manpower." *World Journal of Psychosynthesis* 3:4–5.
3. Richard L. Hatton, "The Family Doctor and TV-Linker Family Practice." *World Journal of Psychosynthesis* 3:33.
4. H. C. Tien, "Medical Television Techniques: Telefusion Couch Therapy." *World Journal of Psychosynthesis* 4:37–39.
5. H. C. Tien, "Psychosynthesis: TV-Linked Cybernetic System of Psychotherapy." In *Progress of Cybernetics,* Vol. 1. J. Rose, ed. (Proceedings of the First Internation Congress of Cybernetics, London,

1969. Gordon and Breach Science Publishers, 1970).
6. H. C. Tien, "Psychosynthesis: Cybernetic Telefusion of Modern Psychiatry" (Scientific Exhibit. Alliance for Engineering in Medicine and Biology, Annual Conference, Washington, D.C., November 1970).
7. H. C. Tien, "Psychosynthesis: A Cybernetic System of Psychotherapy" (Sixth Annual Rocky Mountain Bioengineering Symposium, Laramie, Wyoming, 1969).
8. H. C. Tien, "Psychosynthesis: A TV-Cybernetic Hologram Model." in *The Nervous System and Electric Currents*, Vol. 2. Norman L. Wulfsohn and Anthony Sances, Jr., eds. (New York: Plenum Press, 1971).

2. TV-MONITORED MEDICAL FORMS

The second chapter may appear repetitious to you. It was planned that way, so as to show how the videotechniques are actually carried out with the TV-monitored Forms, 1 and 2; 3, 4, and 5; 6, 7, and 8, so that the reader can eventually duplicate the videosystem in order to obtain a complete history and examination. (See sample at the end of Chapter 4.)

Meanwhile, we are going to repeat the comments of the first chapter so we become thoroughly familiar with the terminology, the theoretical points, and the techniques involved.

The POPF Model (acronym for Professional contact-On television-Professional contact-Feedback) is derived from the PERF Model for TV-linked psychotherapy. (See Glossary.) The POPF Model can be applied not only to psychotherapeutic sessions but also to any learning situation, whether between doctor-patient, counselor-client, teacher-student, family lawyer-family member, when personal contact can be meaningfully amplified by sandwiching TV tapes between personal contacts with proper professional feedback.

In other words, to give a concrete example, the psychiatrist can see the patient in person at the beginning of a TV-linked therapeutic session, to elicit the chief complaint or problems that day; then the patient is left with the same psychiatrist on TV to amplify the therapeutic information. At the completion of a half-hour TV treatment program, the psychiatrist returns for further personal contact and relevant transaction regarding future appointments, medication, testing, etc., so that the loop is completed with feedback. The POPF Model can also be applied to couch therapy, family therapy, interview psychotherapy, hypnotherapy, etc. The general implication of the POPF Model is that one physician can become two physicians, or more, so that it may be an answer to the medical manpower shortage.

By using the POPF Model, we are not here eliminating the physician but in fact intensifying the personal contact. The physician introduces himself to the patient on television and the television personality of the physician begins the TV-monitored medical history. The TV-monitored history is then filled in by the patient, who answers item by item as the physician or psychiatrist obtains the detailed complete medical background of the patient before instituting therapy.

The TV-monitored medical history is, of course, a great timesaver. Moreover, the physician is always even, present, and able to go through all the necessary details to obtain a good analysis of the patient's past in the service of his present needs.

The following dialogue is excerpted from the full television tape that is used to obtain medical histories. In the complete tape, each question is read aloud, with time allowed for the patient's responses, which are written on a printed form.

INITIAL HISTORY AND EXAMINATION

DOCTOR: What did you want to tell me about your problem?
PATIENT: It's hard to explain, I guess. I don't know what to say.
DOCTOR: It's hard to explain? I will help you a little bit. I will go over this page with you very carefully. This is the initial interview, your first session, to learn more about you.
DOCTOR: What is the chief complaint? What would you say, just a few words.
PATIENT: Something's wrong with me.
DOCTOR: Okay, put down "something's wrong with me." If you have any symptoms or signs, like crying spells, put that down. Now, you see under "B," how long have you been concerned about this, approximately?

PATIENT: I thought about going to a psychiatrist a long time ago. Maybe six years ago.
DOCTOR: What really precipitated your decision to do something? What gave you the final decision that now is the time you should do it? This may have been going on for weeks, or months, or years, but you have decided now is the time.
PATIENT: I didn't think my own opinions would be able to help, and I want to get things straightened out before school.
DOCTOR: Are you ready to go on to the next point? We are going to go in deeper yet, and things will become clearer to you. We are going to try to explain the chief complaint. Now, by "A" here we mean, when did it emerge in your mind that you wanted to see a psychiatrist or professional person? Do you remember any incident connected with it? Just some clues.
PATIENT: About two weeks ago I became really positive because I don't think I can make decisions myself, can't grow up.
DOCTOR: In "B" we ask, "Perhaps you can't be sure of the problem, but what really bothers you?" What we mean by that is, how it affects you, bodily, physically, emotionally, or family-wise. How does it disturb you or make you uneasy?
PATIENT: It just pops up out of nowhere. Things can be fine, then I start feeling fear.
DOCTOR: How does it bother you?
PATIENT: Makes me nervous, can't eat, sick to my stomach.
DOCTOR: All right, "E: Is it emotional?" How did you answer that? If you answer Yes, then describe briefly what emotions are involved. Emotions are anger, fear, despair, jealousy, temper, happiness, depression. How are they involved with you?
PATIENT: Fear, guilt, jealousy, mistrust.
DOCTOR: In the next question, "F: Is it mental?" We mean making decisions, can you think problems out? Are ideas clear to you or do you have ideas you don't understand? Thoughts that puzzle you? Not emotional, not feeling, but thinking, ideas.
PATIENT: Much indecision, two sides always pulling.
DOCTOR: Next, is it connected with your family in any way?
PATIENT: What if you think it is partly?
DOCTOR: Just mark Yes and write below, Partly. Who is this family member involved? Write down the person's name and how.
PATIENT: Partly. My sister, what happened to her with getting a divorce, getting sick, that upsets me.
DOCTOR: Now, J is a very important question: "Have you had a similar situation before?" Yes, No, Don't know. How did you answer?
PATIENT: Yes.
DOCTOR: Then put down a short description, about the time or the situation. Every case is different. Just whatever applies to you.
PATIENT: It's vague.
DOCTOR: Now, under O, little b, we are asking: "Do you wish to have emotional stability?" Is your desire in coming here to improve your emo-

tional stability, to handle your emotions? Yes, No, Don't know. If you answer Yes, state which emotions you want to be able to control: love, hate, anger, happiness, fear, depression—what is the emotion you wish to be able to handle? Which emotional things do you wish to be able to handle better? This is a very interesting question. People go to school to improve their mind, to learn to read and write, but very few people are being taught how to control their emotions. And this is what we are interested in, if this is your area.

PATIENT: Yes. Fear, love, happiness, depression.

DOCTOR: Or is your wish to have peace of mind? Whatever it means to you, harmony and smoothness in your life, not get riled up—whatever peace of mind means to you. Everybody, again, is different. If Yes, how do you wish your peace of mind improved?

PATIENT: Yes. Self-fulfillment, growth.

DOCTOR: Do you wish happiness? Some people equate happiness with pleasure. They are not exactly the same. So if you answer Yes, could you clarify what you mean by happiness? How would you know you are happy? Some people say they are so happy, they cried. What is happiness for you? Maybe you have seen the cartoons that say, "Happiness is lying in the grass, happiness is taking a shower, happiness is walking about." What is happiness to you in life? For some people, happiness is to do nothing.

PATIENT: Yes. Living life and not just existing.

DOCTOR: Next question is, "Do you wish to improve your relationship in a family?" This could also mean do you want to build a new family relationship. Like a marriage, or improving your relationship with some person important to you. Do you wish to have a more stable, happier relationship but don't know how to do it? Yes, No, or Don't know.

PATIENT: Want to be myself even with people I don't know.

DOCTOR: Next, did you have any idea when you came to me what kind of therapy, what techniques, how to get there? Yes or No. Do you have any ideas? If Yes, what ideas do you have? There are many ways we can help you, but we must find the best way to fit your situation. Would you circle your preference? If you don't understand, I will go over it with you briefly.

One, psychotherapy, are you interested in that? Psychotherapy—psyche means mind; therapy means service. In other words, we offer some service to you, trying to bring out your thoughts, ideas, feelings, and to figure out exactly what can be improved. Do you have a car? The car sometimes doesn't function. Sometimes it doesn't move for you, sometimes it makes noise, or whatever it is, then you need service. The automobile mechanic would listen, find out, test, and then say, "Ah, there's the problem." Sometimes it is very minor, sometimes it may need a little extra work. But usually he can fix it. Usually. And the mind can be thought of in this way.

Two, couch analysis. Have you ever heard of that? In psychotherapy we are usually face-to-face, like we are now. In couch therapy you do not look at me. If you wish to laugh, cry, swear, say anything you like, we'll bring everything out without the social strain. Then we can get exactly what is on your mind.

Three, drug therapy or medication. Maybe you are just interested in getting some medication to help you with all those things. We have various kinds of medication. Are you interested in that? I do not mean for you to treat yourself. What I am saying is that people do come in sometimes and say, "Doctor, I need an anti-depressant," or "Doctor, I need something for controlling my weight," or "Doctor, I am nervous, give me a tranquilizer." This gives me a clue. That's what I am interested in.

Four, family therapy. What do you think about family therapy? In family therapy you bring the significant individual with you, be it your wife, husband, mother, father, daughter, relative, or boyfriend, whatever. If you are interested, put that down.

Five, electrotherapy. Any kind of electrotherapy, if you have heard about it. I am not offering any of this to you, I am only trying to learn if you have ever heard of anything like this.

Six, group therapy. Many people have heard of group therapy, or perhaps seen it on television. Have you been to group therapy? Are you interested in anything like that?

Seven, personality development, are you interested in that? By that, I mean do you want to develop a higher personality, a better personality? Not that the one you have is bad or anything, but some people want a better one. Just like you may want a better education. A person may be very good, a high school graduate, but he may want to go higher in education. The same thing applies here. If so, circle that.

PATIENT: Psychotherapy; couch analysis; drug therapy or medicine.

DOCTOR: Do you have anything else you are looking for? Do you feel we have covered everything? There is a blank page there and you can write down anything you feel may be of importance to me we have not covered. Have we missed anything? If we have, put it down. Well, we've gone through quite a bit of territory. Write down if you think of anything else.

PATIENT: I would like to talk about drugs and how they have affected my thinking and my thoughts about myself.

DISCUSSION

As you can see in the above, we have now obtained a summary of the medical history of this patient, with the aid of TV monitoring. This TV-monitor approach enables the physician to take several medical histories within the same hour, at the same time he is thorough and complete. In fact, the TV system is adaptable to all physicians, and each medical history would be done as the physician is what he is at his best, when he is prepared, when he is not fatigued, and is willing and consistent in his medical procedures. Since the physician always performs his best, which is the one that has been recorded for future TV-monitored patient histories, then we would have a consistent product and result as needed.

Of course, the TV-monitored history is not restricted to the chief complaint and evaluation of current problem, but it can also go into the past medical history, family history, systemic review, etc. It is the purpose of this series of articles to show how completely the television monitor can duplicate the personality of the physician, to make his medical knowledge and service available to all those who need medical care—as a right and not as a privilege.

MICHIGAN INSTITUTE OF PSYCHOSYNTHESIS

TV-MONITORED FORM

NAME________________ Date________

1. INITIAL HISTORY AND EXAMINATION

0. INFORMANT: Are you answering or replying for yourself? Yes______ No______ If no, for whom, Wife______
Husband______ Mother______ Father______ Other______.

1. CHIEF COMPLAINT: A. What is the principal or central reason you are coming to me?
B. How long have you been concerned about this? Approximate date____________.

2. PRESENT DISTURBANCE: (Clarification of Chief Complaint)
A. What concerns you most at this point?
B. What other problems do you wish to bring up?
C. What else really bothers you?
D. What made you decide to seek professional opinion or help at this time?

Explanation of Chief Complaint:
A. When did you first begin to have this concern or problem?
B. Perhaps you can't be sure of the problem, but what really bothers you?
C. How does it bother you?
D. Do you feel your problem is physical? Yes______ No______ Don't know______
If Yes, describe briefly.
E. Or is it emotional? Yes______ No______ Don't know______. If Yes, describe briefly.
F. Or is it mental? Yes______ No______ Don't know______. If Yes, describe briefly.
G. Or is it family? Yes______ No______ Don't know______. If Yes, describe briefly.

H. Or is it marital? Yes______ No______ Don't know______. If Yes, describe briefly.

I. Or is it a divorce problem? Yes______ No______ Don't know______.

If Yes, do you want to save your marriage? How?

J. Have you had a similar problem before? Yes______ No______ Don't know______.

K. Is it a school problem? Yes______ No______ Don't know______. If Yes, describe briefly.

L. Is it a sexual problem? Yes______ No______ Don't know______. If Yes, describe.

M. Is it a work or employment problem? Yes______ No______ Don't know. If Yes, describe.

N. Do you have financial or money problems? Yes______ No______ Don't know______.

If Yes, give figures, income, debts, etc.

O. Now that you have come to see me, what do you hope to achieve?

a. Do you want to improve your physical health? Yes______ No______ Don't know______.

If Yes, name the physical problem.

b. Do you wish to have emotional stability? Yes______ No______ Don't know______. If Yes, state.

c. Do you wish to have peace of mind? Yes______ No______ Don't know______. If Yes, describe.

d. Do you wish to gain happiness? Yes______ No______ Don't know______.

If Yes, clarify what is happiness to you.

e. Do you wish to improve family stability? Yes______ No______ Don't know______.

If Yes, what areas?

f. Do you wish to improve your marriage? Yes______ No______ Don't know______. If Yes, what area?

g. Did you have any ideas of what kind of therapy you are seeking? Yes______ No______.

If Yes, describe, or circle your preference:

1. Psychotherapy. 2. Couch therapy analysis. 3. Drug therapy or medicine.

4. Family therapy. 5. Electrotherapy (i.e., ELT).

6. Group therapy.

7. Personality development. 8. Other____________.

(If you need more space, use back of page.)

3. TV-MONITORED MEDICAL FORMS

INTRODUCTION

This is the second in a series of TV-monitored interviews. The techniques can be used not only in psychiatry and family practice; they are applicable for forensic psychiatry or legal work in general to get a systematic evaluation of past history and to be transformed into a readable format.

In order to make use of the present system, we have already introduced Forms 1 and 2 in the last chapter of the manual. This is a continuation to determine the past problems that may have bearing upon the present complaint or present conflict or crisis. And we are also trying to determine what previous treatment the person or family may have had or they may be looking or seeking for.

The third form in this series, called personal history, is used to try to determine from conception to the current development of the personality as to his political and philosophical views of life, his upbringing, education, religion, and so on. (The forms are reproduced in the Appendices.)

These forms are entirely TV-monitored and perhaps it would be worthwhile to emphasize this point. This TV-monitored program can be adjusted and can be corrected through feedback process at a different speed, rate, or educational background applicable to the patients or families. We are currently only illustrating how this TV-monitored interview can be done. There appears to be no limit as to how elaborate or complex or detailed this can be developed.

The following is an illustration of how they have been done:

DOCTOR: I am a family psychiatrist and our job is to find out all the problems you may have, whether they be physical or medical or family or emotional, or at work, and try to get some understanding in your situation. So let me ask you (to son), did you make the appointment?

SON: Yes.

DOCTOR: Why was that?

SON: Because my mother was having a lot of headaches and she couldn't get rid of them. Another reason I made it was because you'd be talking to her or something like that and it would be an off-the-wall conversation.

DOCTOR: What kind of conversation?
SON: Well, it wouldn't be related to what you were talking to her about. She changed the subject.
DOCTOR: And you are worried about that? You are living at home?
SON: No.
DOCTOR: Where is your dad?
SON: He is at work, but he lives at home.
DOCTOR: How many children are there?
SON: 11.
PATIENT: 12.
DOCTOR: Well, you are remarkable. Maybe you need some help, but you deserve some help! Now what we need to do is go into some details, and here is paper for you to write them down. We go into the past problems to bring the present into view. Write down your name and date there. I don't help you with the date, to give me an idea of your memory, how alert you are. Put the year down, too. Would you read the first line for me?
PATIENT: "1. Have you had the same or similar problems before?"
DOCTOR: Have you had the same or similar problems before, like headaches, or whatever trouble you may have.
SON: She's been having headaches, but now they are getting worse.
DOCTOR: Give me the approximate dates, when you first started to have symptoms. Can you describe the symptom, too, a little bit? Was it very severe? Did you have to quit work? Were there associated symptoms? Next question, number 2, have you been to any other psychiatrist, a family psychiatrist or a psychiatrist of any kind? Yes or No?
PATIENT: You mean for myself?
DOCTOR: Or for anybody.
PATIENT: Well, I talked to one concerning my daughter.
DOCTOR: Would you put that down, and write your daughter's name down, so I have some idea. What was the problem about? That may give me a clue what you worry about. The daughter worried the mother.
SON: Right now she isn't home. My mother is taking care of all her kids plus her own, and the daughter can't be found. That worries her, too.
DOCTOR: Number 3, what problems did you take to other doctors? That is just any doctor you can think of.
PATIENT: Well, abdomen pains.
DOCTOR: Abdominal pains, in your belly? Okay, put that down. Would you read number 4 for me, please?
PATIENT: "Why didn't you continue to work with these doctors? Give three reasons."
DOCTOR: Yes, why didn't you continue to work with these doctors? We know that to solve problems requires time and effort and continuity. Why didn't you continue to go?
SON: Did they stop hurting?
PATIENT: Because I began to seek for a deeper depth, well, more in the spiritual vein.

DOCTOR: So you were looking for a different way out, a spiritual way? Any other reason?
PATIENT: No.
DOCTOR: Okay, read the next one.
PATIENT: "5. Do you see any connection between your present problem and your past?"
DOCTOR: What is the connection?
PATIENT: Well, that's a spiritual problem, too.
DOCTOR: Just put down a few words or a sentence, then we will put things together. The idea in our evaluation is to cover as much ground as possible, in the shortest possible time and most efficiently and economically. Because you are a certain age right now, you have lived so many years. For us to get the whole picture, we've got to put it all in perspective.
SON: Yes, because this idea could have been troubling her for quite awhile, but is just beginning to surface.
DOCTOR: Yes, even in childhood. Next question.
PATIENT: "6. Is there anything else you wish to say about past or present problems?"
DOCTOR: Put something down if any comments come to your mind. Anything you put down is good.
SON: In the past, you were easy to get along with. Now with your present problems, you are not too easy. You know what I thought it was? Changes of life.
DOCTOR: Change of life. You can put that down if it applies. You know, teenagers have teenage problems and mature women have mature problems. The important thing we have learned, as I mentioned, is that anything you put down is good. Because families usually tell the truth. I have yet to see a family that doesn't tell the truth. If you have finished that, we will do the next page. The more we can do together, with the television to help us, the more we will be able to help you.

PREVIOUS TREATMENT

DOCTOR: Be sure to answer every question; it will help to bring your problem in focus. It's like a doctor looking into a microscope. He's got to put it into focus to see what the tissue is. In your case, it is what your family or physical or emotional problems are. We are trying to zero in, get up close. Number one, would you look at that?
PATIENT: "1. Have you ever had counseling?"
DOCTOR: By that we mean if you say you are looking for spiritual help, you could have seen a minister or psychologists, social worker, welfare people, whatever. Have you ever had counseling of any kind, school, work, anywhere? Yes or No.
PATIENT: School.
DOCTOR: From whom did you get this?
PATIENT: Over to the high school.

DOCTOR: Put that down. What was it about?
PATIENT: English and so on.
DOCTOR: I see. Learning. Okay, read the next question.
PATIENT: "2. Have you ever been to a mental health clinic?"
DOCTOR: Have you ever received help from a mental health clinic, yes or no?
PATIENT: Not for myself.
DOCTOR: Next question.
PATIENT: "Have you ever been to a psychiatric hospital?" No. Well, let me see. My daughter was in a state hospital, so I went there.
DOCTOR: You can check that, for your daughter, then. Anything you put down is good. It's when you don't tell me anything, then of course I don't know. You tell me anything, I will be able to figure it out a little more. All right, we will continue. Have you ever had therapy of any kind for your problems, yes or no?
PATIENT: No.
DOCTOR: Number 5.
PATIENT: "Have you ever had nerve pills?" Yes.
DOCTOR: There is a chart to list them. Do you know the name? Then just put down the color and the shape if you don't.
SON: What about the ones you get out of the drugstore?
DOCTOR: Sure. Self-medication is of interest to me.
SON: You take a lot of aspirin, don't you?
PATIENT: No.
SON: Not any more? A lot of coffee, too. Coffee has something to do with that, too, doesn't it? She makes it so strong it will dye the cup.
DOCTOR: We will find out why she needs these things. Number 6, have you ever had any electrotherapy of any kind?
PATIENT: No.
DOCTOR: Have you anything else to add, medical or otherwise? By that we mean, if you have had group therapy, or family therapy, or electrotherapy, or couch therapy, put that down. I will give you time to put something down and then we will go on to the next one.

PERSONAL HISTORY

DOCTOR: Now we will go in a little deeper with the next one. The first thing to answer is about conception. That means how you were conceived, did your mother carry you normally, yes or no?
PATIENT: No.
DOCTOR: It was not normal? What was the trouble?
PATIENT: Premature.
DOCTOR: Then you write down "premature." That's good. How many months' baby?
PATIENT: Seven.
DOCTOR: Now I have learned something about how you came into the world. The next question, "Were you wanted, yes or no?"

PATIENT: Yes.
DOCTOR: Write down Yes. Were you conceived out of wedlock?
PATIENT: Yes.
DOCTOR: Write that down. So you were the first-born? We learned something else about the background. You see everything happens for a reason, that's why we go into detail. Perhaps this you have never even heard of, it's new to you (to son).
SON: That's right.
DOCTOR: You are a little bit surprised.
SON: I didn't know some of this.
DOCTOR: Now you can see how much your mother has been carrying all these years, burdens of spiritual problems.
SON: She told me I wouldn't understand.
DOCTOR: Next, was your birth normal, yes or no? The birth process, that's all it means. You were not full-term, so check that one. Were you delivered in a hospital?
PATIENT: No.
DOCTOR: Besides being a preemie, did you have any complications?
PATIENT: Yes. Pneumonia.
DOCTOR: Some people may not have any problems. Each case is different. Put down what applies to you. How about birth injury?
PATIENT: No.
DOCTOR: Then we go on to school. Some people go to nursery school, others don't. What is your case, did you go to nursery school?
PATIENT: No.
DOCTOR: Most people go to grade school, start at kindergarten. What age did you start?
PATIENT: Six.
DOCTOR: Okay, put that down. What kinds of grades did you get in grade school? Would you say you were an A student, B, C, D, or E?
PATIENT: B and C. Average student.
DOCTOR: Same thing applies to high school. How many years did you go to high school? If you have gone, fine; if not, also fine.
PATIENT: Junior high.
DOCTOR: A couple of years, okay. What kind of student were you: A, B, C, D?
PATIENT: Average C student.
DOCTOR: If you have gone to college, fine, do the next one. If not, just omit it. The next step is dating. How old were you when you started to date?
SON: It's been a long time, hasn't it?
PATIENT: It's kind of hard to remember.
DOCTOR: Just give me anything that comes to your mind. It is sometimes better that way.
PATIENT: About thirteen, I'd say. As close as I can remember.
DOCTOR: The next part is, was it steady dating?
PATIENT: No, not really.
DOCTOR: When did you get steady, with whom? Put that down. How

long did you date? We just get a capsule look. And also put down the first important emotional relationship, commonly called falling in love. Okay, who was it? And what happened to that relationship? Sometimes you marry the first love.

PATIENT: We were married and lived together about a year.

DOCTOR: No children?

PATIENT: We had one child. Then we separated.

DOCTOR: People often do the same thing. They hate to admit it, but they call it infatuation. It is probably always the most important experience.

SON: That's when you fall so hard.

DOCTOR: What kind of wedding did you have? How old were you when you got married?

PATIENT: Sixteen.

DOCTOR: What kind of wedding did you have, by a judge, or justice of the peace, or captain on a boat, or minister or rabbi?

PATIENT: Minister.

DOCTOR: You were divorced once. Put down whatever applies there. Then we get into employment a little bit. What was your first job?

PATIENT: Housework.

DOCTOR: Then think about the first important job you had, where you really felt it was a real job, that was important to you.

PATIENT: In a department store.

DOCTOR: Put down how much you made, too.

PATIENT: About thirty-five dollars a week.

DOCTOR: So that gives me some idea how far you have come economically. Then we will go on to religious background. What church do you belong to and what do you practice? Some people may be brought up in one religion and then become something else.

PATIENT: Well, my religion varies. It is mostly missionary work.

DOCTOR: What kind of missionary work?

PATIENT: I would rather leave it at missionary work.

DOCTOR: Now the question on political views. Give me some idea how you struggle with the world and the world struggles with you. You can put down anything you like, really—conservative, liberal, whatever you think.

PATIENT: I will just leave it at people don't understand me.

DOCTOR: Now your philosophy, how do you look upon life, what is life all about?

PATIENT: Life is thinking of others. Not from a selfish point of view.

DOCTOR: Then we go into outlook on life. Do you think life is a great future, no future, absurd, interesting, what do you think? Are you a pessimist, optimist, realist? If you are a pessimist, treatment becomes difficult, because you feel treatment is going to be a failure.

PATIENT: Life always has a view. You don't think of life just as a standstill. You continue to grow. It's a daily growth.

DOCTOR: Very good. I think the chances of working together, getting somewhere, are good. Now your self-description. What kind of a person do you think you are? Who are you, really? Just put down a short sentence.

PATIENT: I'm a child of the king.

DOCTOR: Then take all the space you need to make any other comments.

DISCUSSION

The first point to discuss is the view that personal contact is reduced by the use of television with the patient. Indeed, personal contact is reduced, but not as much as one realizes if we follow the POPF Model. The POPF Model (acronym for Professional contact-On television-Professional contact-Feedback) is derived from the PERF Model for TV-linked psychotherapy. (See Glossary, page 167.) In this model, which can be applied not only to psychotherapeutic sessions but also to any learning situation, the psychiatrist can see the patient in person at the beginning of a TV-linked therapeutic session, to elicit the chief complaint or problems; then the patient is left with the same psychiatrist on TV to amplify the therapeutic information. At the completion of a half-hour treatment TV program, the psychiatrist returns for further personal contact and relevant transaction regarding future appointments, medication, testing, etc., so that the loop is completed with feedback.

This way we are actually amplifying the personal contact with the thru-put information, where the input and output are determined by personal contact. In other words, the thru-put is the amplified system through television, where input and output are still determined by professional contact in this approach. Furthermore, we must look upon the larger picture. If the physician, psychiatrist, or lawyer does not have time to see families at all, how much more impersonal the world would be.

With the whole approach we have made possible, both economically and time-wise, the evaluation of multiple family members in the same session. For example, if a couple comes and the wife points her finger at her husband, saying, "He needs evaluation," and the husband says, "Oh, no, she is the sick one," we can settle the issue right away by modern technology. Both can be evaluated simultaneously and impartially by TV-monitored approach, while the professional person would still have time to do more objective analysis, thinking and assessment when the data are in.

Second, the TV-monitored approach is helpful to those people who cannot read or write. They can listen to the questions on tape, respond verbally, and their answers on tape can be transcribed by a secretary. A TV programmer can be present to assist the family which needs help because of educational or physical limitations.

The third point to discuss is the idea of a TV programmer. The TV programmer need not have very elaborate education. A person with a high school education and sufficient preparation and some training would be able to become a very effective paramedical, parapsychiatric, or paralegal assistant in the TV-monitored approach.

Finally, this second in a series points to the beginning of a medical specialty. This new specialty may well be described by the term "videology." Videology, as has been developed at Michigan Institute of Psychosynthesis, has definite implications regarding medical education and the training of the future physician.

For the sake of symmetry and completeness, the other professionals, e.g., law or education, the term "videoist" can be reserved for those who use television technology and the term "videologist" would be reserved for use by physicians and medical scientists.

SUMMARY

This is the second in a series of articles on TV-monitored interview techniques. Such techniques are adaptable for obtaining medical history, background and evaluation of family problems, and they may be applicable to other professions besides medicine. The concept of videology is introduced and the terms videologist and videoist are defined. A third article is in the next chapter.

4. TV-MONITORED MEDICAL FORMS

INTRODUCTION

This is the third of a series of TV-monitored techniques. As ideas and techniques evolve, we have come to use the compact term "videology" to include all TV-linked techniques. At this point, perhaps we should list the types of TV-linked sessions used in videology: 1. The Primary Session; 2. The Ordinary Session; 3. The Booster Session.

1. *The Primary Session*: The professional, physician or therapist, is live on the television monitor. The primary session helps to record not only the patient's responses but also the physician's approach.

2. *The Ordinary Session*: The session is sandwiched between professional contact, in terms of the POPF Model. (In POPF Model, acronym for Professional contact-On television-Professional contact-Feedback, for example, the physician or therapist can see the patient in person at the beginning of a session, to elicit the chief complaint or problems; then the patient is left with the same professional on TV to amplify the therapeutic information. At the completion of the TV treatment program, the professional returns for further personal contact and relevant transaction regarding future appointment, medication, testing, etc., so that the loop is completed with feedback.) Of course, the ordinary session gives us the power of amplification of information, analogous to a microscope. We have lens at both ends in order to amplify space, and in TV-monitored sessions we have information at both ends to amplify time, thereby increasing information flow from professional to those who are being served.

3. *TV Booster Session*: The entire session is TV-monitored without the professional's presence. Medical history can thus be obtained at minimum cost and diagnostic and therapeutic information can be economically transferred.

The three types of sessions, primary, ordinary, and booster, are basic and can be diversified in many combinations to meet the patient's needs and the physician's needs, in order to provide optimum service in terms of knowledge, time, and economy. For example, a TV booster evaluation can be achieved with a medical history at one-third the expense of the primary time. The demand for primary sessions may not always be necessary, but may result from tradition, conservatism, or habit, rather than the actual need of the situation. It is therefore a matter of education. Videology is yet at its infancy, but its growth will be rapid since the demand for service is a persistent one. The third of this series introduces how we can obtain the medical history through videology, so that the time demand on the professional is lessened and the economic burden on the family is reduced. The procedure described here can be used for primary, ordinary, or booster situations, as demanded by the individual physician and the special needs in a given situation of the patient. (The forms used are reproduced in the Appendices.)

FAMILY HISTORY

DOCTOR: What I would like to go into is your family background a little bit. Would you read the top of the page?
PATIENT: "Please fill out the following form as completely as possible. List the first names of your father, mother, brothers and sisters, in chronological order and supply the following information about each."

DOCTOR: This is family history. We are going behind the scenes a bit. In the first column, we ask the name, age, education, marital status, where the person is living, and life adjustment, and any comments you wish to make. Put down the name of your natural father. If he is living, write down "L." Put down his age. What does he do for a living, predominantly? Is he married now? Was he married before, or divorced? Where is he living now? How would you rate his adjustment, good, fair, poor? Under comments, just write anything, any remarks you wish, to give me a clue about his personality or your relationship to your father. You don't have a stepfather or adopted father?
PATIENT: No.
DOCTOR: How about your mother? Living? What age? List her education, occupation—that is, what she considers herself to be, whether it is housewife, teacher, or whatever. If she was married before, you can note that. How is her adjustment, good, fair, or poor?
PATIENT: Good.
DOCTOR: Is there any comment you want to make, anything about her personality relative to you or your relationship? You don't have a stepmother or adopting mother?
PATIENT: No.
DOCTOR: Now we want the same information for your siblings. Do you know what that means? Brothers, sisters, how many do you have?
PATIENT: One brother and one sister.
DOCTOR: Put their names down. Are they living? Age, education, occupation, as it applies to each one. Some families have half-brothers and stepsisters, and so on. It can be quite involved. After we finish the siblings, then we can go on to your spouse, if you are married.
PATIENT: What if my brother got a divorce and then remarried, should I just list him as married?
DOCTOR: Yes, but also put down the divorce or make a comment. We

can explore that later on, how you reacted to it. Have you finished the part on siblings? Next we will fill this out for spouse. You are married, right? Her name, age, education, occupation, marital status. Was she ever married before? Then any children or ex-spouses. You yourself have never been married before, have you?
PATIENT: No.

SYSTEMIC REVIEW

DOCTOR: This next form is a physical review. We will go over it together to make sure everything is clear. How would you describe your complexion? People use all kinds of terms.
PATIENT: Average.
DOCTOR: How about body build, what would you say?
PATIENT: Medium.
DOCTOR: Then your weight, height, hair color, eye color. Have you ever been seriously ill?
PATIENT: It depends on what you mean. Except for tonsils, I have never been in the hospital for anything. I had strep throat once, with a high fever.
DOCTOR: Next we ask when was your last medical examination? By that I mean a real medical examination, covering all your systems, not just a short check-up.
PATIENT: I have no idea. I did have a check-up before I came here to school.
DOCTOR: Okay, put that down and the doctor's name if you remember it. Next, to what extent have you used alcoholic beverages, beer, whiskey, hard drinks, what have you?
PATIENT: Not very much.
DOCTOR: The best way we can gauge sometimes is to put down how much you spend a week for alcoholic drinks, a dollar, two dollars, five dollars, etc.
PATIENT: A week? Less than a dollar.
DOCTOR: Next, do you use or have you ever used any narcotic drugs? Today we broaden that out to include controversial drugs like marijuana, LSD, mescaline, speed, goofballs. We can explore in the future why you used them.
PATIENT: Yes.
DOCTOR: Have you taken treatment for drug or liquor habit?
PATIENT: No.
DOCTOR: Number 4, previous hospitalization. By that we mean have you ever been in the hospital overnight? Date, nature of the hospitalization, who was the doctor? Have you had other hospitalizations or operations, whether as an inpatient or an outpatient? Next, previous trauma or accident. Is the test tedious for you?
PATIENT: No.
DOCTOR: Sometimes it is very detailed. Now we are on Number 7, al-

lergies. Are you allergic to any food, drug, or substance?
PATIENT: Not that I know of.
DOCTOR: Number 8, eyes. Do you know what you wear glasses for?
PATIENT: Near-sighted.
DOCTOR: Any other trouble with your eyes? Next, ears. Is there any history of deafness, any impairment, discharge from the ears? How is your hearing?
PATIENT: All right.
DOCTOR: Okay, Number 10. Nose trouble, nosebleeds.
PATIENT: No.
DOCTOR: Your neck. Do you have a goiter? Enlarged thyroid? Any other glandular trouble?
PATIENT: No.
DOCTOR: Number 12. Heart trouble. Have you had any history of heart attack or disease, high blood pressure? Any rheumatic heart disease, angina pain? Number 13, lungs. Have you ever had TB, pneumonia, pneumonitis?
PATIENT: No.
DOCTOR: Next, any stomach trouble, history of vomiting, stomach ache? Intestines, any diarrhea?
PATIENT: Not very often, occasionally.
DOCTOR: Put that down, so we can explore why you have this. Do you have any idea?
PATIENT: I usually get it when I eat any kind of meat, beef.
DOCTOR: Constipation, do you use anything for that?
PATIENT: Very seldom.
DOCTOR: Next, any gall bladder trouble, liver disease, jaundice or turning a yellow color? Number 16, urinary disease—your kidney and bladder. Any blood in your urine?
PATIENT: No.
DOCTOR: Any history of syphilis, gonorrhea? Number 17, have you ever had any glandular disease other than what we have mentioned? How about sugar diabetes, hyperthyroidism, Addison's disease?
PATIENT: No.
DOCTOR: Have you ever had any nerve disorder?
PATIENT: No.
DOCTOR: Have you ever had any nerve disorder?
PATIENT: I don't know. For awhile, I thought maybe that was what was giving me headaches.
DOCTOR: Put down how many years you have had these headaches. Also, write down any head injury, head trauma, if you were ever knocked out or had convulsions or spells, loss of consciousness, passed out, fainting.
PATIENT: Not that I can recall.
DOCTOR: Number 19: Have any of your parents, brothers, sisters or near relatives ever had diabetes, mental illness, or tuberculosis?
PATIENT: My father had three nervous breakdowns.
DOCTOR: Put that down. We will explore what you mean by three

nervous breakdowns. How old was he?
PATIENT: He had one in 1950, one in 1954, and the last one was in 1968.
DOCTOR: This may give me some clue to your situation. Now we will go on to Number 20. How long have you been at your present weight?
PATIENT: About three months.
DOCTOR: What caused the change?
PATIENT: I haven't been eating properly.
DOCTOR: Put down how much weight you have lost. Number 21, are you on any medications at the present time, or any drugs, any controversial drugs? If you are still on them, please put that down, whatever you are taking. Diet pills, vitamins.
PATIENT: No prescription drugs but I do smoke grass.
DOCTOR: Now the next part of the form applies to women, not to you. Do you know your wife's menstrual cycle? Is she pretty regular?
PATIENT: Yes.
DOCTOR: Is she on contraception?
PATIENT: Yes.
DOCTOR: She doesn't have any serious problems? Her menstrual period does not affect you?
PATIENT: No.

TREND AND CONTENT QUESTIONNAIRE

DOCTOR: Now to the next form. Trend and Content Questionnaire. Again, put your name and date there so we can keep track. Number 1, what do you worry about most of the time? Everybody worries about something, so what do you worry about? Put down whatever comes to your mind; reject nothing.
PATIENT: Money and the military draft.
DOCTOR: What do you fear? Everybody is afraid of something. Even small fears, or fears you feel are ridiculous but you still have them, put them down.
PATIENT: Getting drafted.
DOCTOR: Why don't you read Number 3?
PATIENT: "Anything make you angry?"
DOCTOR: Yes, does anything make you angry, upset, lose your temper, have a tantrum?
PATIENT: All I can say is, yes, some things do.
DOCTOR: Like what, give some examples.
PATIENT: Well, nothing in particular. Nothing that I consistently get upset about.
DOCTOR: Give an example of an occasional reason, then.
PATIENT: I really can't because there is nothing that stands out in my mind.
DOCTOR: Okay, read Number 4.
PATIENT: "Do you have any strong dislikes? Hatred?"

DOCTOR: Dislike of people, things, or situations, even ideas? Does it reach the point of hatred?
PATIENT: No.
DOCTOR: How about reading the next one?
PATIENT: "Does anything or anyone make you jealous? Envious?"
DOCTOR: Does anything or anyone make you jealous that you are aware of? How you tell is, whenever something good happens to somebody, you start criticizing them. This is one way you can tell.
PATIENT: No.
DOCTOR: Envy is not disagreeable. Envy is getting a feeling you are going to share his good fortune. Jealousy means you won't be able to share his good fortune, or whatever the situation is. That is how you tell them apart. You could be envious of your wife because she is going to get a raise but she is going to share part of it with you. You could be jealous of a co-worker because he got a vacation and you might have to do extra work. Okay, read Number 6.
PATIENT: "Does anything make you particularly sad or unhappy?"
DOCTOR: Does anything depress you, make you blue? Do you get unhappy, brood about it?
PATIENT: I will say yes, but again it is nothing in particular.
DOCTOR: Some people answer a different way. Does anything make you happy or joyful? What, for example? Next, I am going to give you three wishes, any wish you want. What three most important wishes would you have in your whole life? Any wish. Absolutely let your imagination take over.
PATIENT: 1. Health for my wife and I; 2. A comfortable life; 3. Wisdom.
DOCTOR: Now after you have looked them over, why do you think they are important to you?
PATIENT: Because I would be happy with these.
DOCTOR: Number 8, besides wishes, do you have dreams at night? Describe one—especially dreams that repeat.
PATIENT: I don't have any.
DOCTOR: Next, why do you think you are now seeking professional help? What is the real reason? Try to be as honest as you possibly can. What are you looking for? I will also be frank with you. Sometimes people come ostensibly for one reason but really they have something else in mind. If I know what is in mind, we can get to it faster. Nothing is too big or too small, or embarrassing, or unreasonable, as long as it is important to you.
PATIENT: Because it often helps to have an outside opinion of one's life.
DOCTOR: Number 11, What are some of your explanations of your problems? Try to name three reasons that come to your mind. Also give me a fantastic reason why things are this way, something out of this world. Three reasonable reasons, one fantastic reason.
PATIENT: I haven't had time or the opportunity to find out everything about myself. I have karma.

DOCTOR: Then we go on to Number 12, What ways have you tried to solve these problems? You must have done something. It is my opinion that anybody who tries to get professional help must have tried on their own for some time.
PATIENT: By facing my faults and trying to correct them, I try to solve my problems.
DOCTOR: Number 13 is left open for you to put down anything you wish that we have not covered.

SUMMARY

This is the third of a series on TV-monitored techniques of videology and definitions are offered for primary, ordinary, and booster TV sessions. The concepts of personal contact and modern technology are discussed. Another article is being prepared in terms of more specific application, besides data gathering, in therapeutic techniques. These articles are signposts to indicate that videology may expand faster than radiology, because it is truly a technology that can go wherever medicine and psychiatry are being practiced.

5. VIDEO-MONITORED MEDICAL HISTORY: FOUR CASES

Television can be harnessed to serve medicine and mental health. Up to now, our lives are dominated by television in a passive way, but television technology can be made into an active therapeutic force in the practice of family medicine and psychiatry. Medical television, videotechniques, and videotherapy can enhance the skills of all physicians. It is especially needed in information processing and feedback. The transfer of information from patients to the doctors and vice versa are fundamental to the healing arts. In order to make health delivery systems more efficient and more economical for the patients and the physicians, videotherapy will serve as a positive contribution. The video image of the family physician or the family psychiatrist can be made available in the various types of clinical situations and also in all types of therapy, whether in general practice, in psychiatry or other medical specialties.

We are all aware that commercial television has been used for entertainment and advertising and oftentimes in questionable TV programs in our society. However, videotherapy is a new development. Videotherapy is not a substitute for the personal contact of the physician or family psychiatrist, but rather as an economical and tireless adjunct of the professional, whatever his specialty. It is apparent that television has been used beneficially in medical schools for educational purposes, as a time-saving technique for the faculty as well as an invaluable tool for learning and review for the students. It has also been used to advantage in recording and demonstrating surgical procedures. The consistent use of television and its powerful videotechniques in family medicine and psychiatry are yet to be realized. To make videotherapy and videotechniques a real part of the practice of medicine, we should apply this technology for the routine work of medical diagnosis and treatment.

Having mentioned the potential areas of application of videotechniques, it is helpful to consider why the development of videology is so vital in the practice of medicine. It has been suggested that physicians, in particular psychiatrists, are still practicing on the level of a cottage industry. Videology and videotechniques can narrow this gap, when electronic technology can make service available to all who need it.

The following reasons are offered to point out why we must apply modern technology to the health challenges in the practice of family medicine:

The emerging practice of family medicine requires special technological advancement in order to meet ever-increasing demands for quality care.

Television technology is very familiar to the public, and as such, it does not require too much introduction or acceptance as a tool in the daily use in the office of any physician.

With the use of technology, the family physician and other specialists can provide more time for direct patient contact to create professional time for intensive psychotherapy as needed.

Routinized and repetitive medical procedures or explanations can be transferred over to the use of videotapes, so that the physician can devote his time to the patient's special and unique problems.

To reduce misunderstanding, videology can assist the family physician with many of the time-consuming activities that he has to perform including the discussion of fees, hospitalization, and other socioeconomic problems of the family.

The video image of the family physician can be a powerful force to maintain his good doctor-patient relationship, whenever the patient is waiting to see the physician in the reception room. The video image of the physician to give information about the doctor's practice that is pertinent to the patient may be given to him during that waiting period. This is good public relations in that no time is wasted in the reception room.

The annoyance and impatience and the proverbial conflict in the reception room can be eliminated or reduced to a minimum when the patient can be sent to a video room to view a program that is relevant to his particular situation, for instance, diabetic management. This, of course, will accelerate good results and improve rapport.

A sound doctor-patient rapport is fundamental to the delivery of good medicine. It is naturally hoped that complaints and grievances will be reduced and eventually malpractice suits could be curtailed from the harried practice of a dedicated physician. Videology means better communication, and better communication means better medical education for the public.

In some of the complex specialties, involving hazardous neurosurgical procedures or orthopedic procedures, videotaping before, during, and after the operation is almost a *sine qua non* for the maintenance of a good doctor-patient relationship. Video recording could help the patient who might have temporarily forgotten the details of his illness prior to the general anesthesia, especially brain surgery. Similarly, in psychiatry, electrotherapy can be videotaped for such an informed consent.

There is a tendency for patients to forget, or for the family members to distort the original reasons for which they sought help. This has been demonstrated experimentally, how general anesthesia could even make the patient forget the details of the previous surgical consultation, leading to the disruption of the doctor-patient relationship and its consequences, despite excellent results.

The above list is only a brief introduction to the various possibilities of videology and videotechniques as they can be applied to the practice of medicine and family psychiatry now. The author is sure that the reader could create and apply other videotechniques in the light of his own special area of clinical or research work.

The following clinical situations will serve as excellent examples of the uses of videology:

1. Initial visits to the physician can be easily recorded for progress and follow-up.

2. Videotechniques can be used for initial family history or mental status examination which can be saved for accurate reference as occasions require in objective documentation.

3. A physician can instruct patients personally, through his own video image, in order to offer his special therapeutic programs or instructions. For example, a physician on video can tirelessly and carefully explain the same procedures regarding diets, contraceptive techniques, prenatal and postnatal care, pediatric instructions, allergy programs, and others as they are developed.

4. Video programs which are pretaped by the physician himself (emphasizing the video image of the physician) tend to be more persuasive in the instructions of his patients.

5. The use of videotapes can resolve time conflicts in scheduling of appointments or family conferences, such as a medical diagnostic review which can later be shown to the absent family member.

6. Videotechniques can be used for preoperative demonstration before the patient is hospitalized, in order to prepare the patient physically and psychologically for the procedure.

7. Video programs can offer either preoperative instructions or postoperative exercises, so that the patient can be active in his own recovery, for example, in physiotherapy.

8. The use of videotechniques to record informed consent and to satisfy medico-legal requirements should be obvious to any practicing physician.

9. Videology can be used not only for the patients, but also for the physicians, especially the family physicians, to assess their own techniques and offer opportunities also for the continuing education of others.

10. Medical histories can be standardized and effectively applied with the use of videotechniques. Videotaped histories, of course, have high medico-legal value.

In summary, it is time to acknowledge that videotechniques and videology must serve as an indispensable therapeutic aid to the practice of medicine. It is also time for all practicing physicians to consider the general use of television technology, not only for medical education by the academician, but also for the day-to-day practice of medicine by the physician.

The following case histories are examples of how videology can be applied to the taking of medical histories.

CASE 1

NAME: A. L.
AGE: 27
SEX: Male
RACE: Caucasian
MARITAL STATUS: Married
EDUCATION: 9th grade
OCCUPATION: Mechanic
BIRTHPLACE: Michigan
IDEOLOGY: Christianity
REFERRED BY: Family physician
INFORMANT: Patient

1. *CHIEF COMPLAINT*: (Given on 5-4-68). "I have been concerned about my personality and my marriage for some time."

2. *PRESENT DISTURBANCE*: According to the patient, "My wife couldn't have children even though we have been married for six years. I have been going out on her a few times. My wife filed for divorce four months ago, but I went out on her even before that. I would tell the other girls I was divorced. I'm not sure whether I want a divorce or not. I have a problem taking care of bills, and sometimes I don't get to work on time or do things on time. I don't know really what I want to do. My job is all right but I don't like it too much. I called a doctor because my sister has gone to him and I thought I had better get some help. I wanted to talk my problems out to see what they are all about. This other doctor couldn't take me right away, but I wanted to do something for myself now so I decided to come to you."

3. *PAST DISTURBANCES*: When patient was asked if he had similar problems before, he claimed, "It's hard to explain. I've had this problem all my life because of the family upbringing." Patient claimed, "Why do I say the hell with everything when I'm drinking or feeling low or when something doesn't go right for me?"

4. *PREVIOUS TREATMENT*: Patient claimed he has had no previous treatment.

5. *PERSONAL HISTORY*: Patient claimed he had a normal conception, wanted, out of wedlock. His birth was normal, in a hospital, with no complications or injuries. He began grade school at age six and his marks as a student were C. He completed through the ninth grade of

high school, with marks of C. He was married at age twenty-one, in a large church ceremony and with the approval of both sides of the family. His first job was at age seventeen, in a store. His first important job was at age twenty, in a garage. His last job was at age twenty-three, as a mechanic. He claims he has no political views. His philosophy is, "Life is short, live it up now; when you are old, you can't." His self-description is, "I am too good at heart, too honest with everything but my wife."

6. *FAMILY HISTORY*: Father, age fifty-seven, eighth grade education; deceased. Mother, age forty-nine, 11th grade education, housewife, remarried, "good" life adjustment. Brother, age twenty-six, high school education, married, life adjustment is "fair." Sister, age seven, student, life adjustment is "fair." Wife, age twenty-five, office worker, life adjustment is "fair."

7. *SYSTEMIC REVIEW*: Patient has a medium body build, is 5′10″ tall and weighs 170 pounds. He has black hair and hazel eyes. He claimed he has been seriously ill before but did not explain. His last medical examination was in February 1958. He claimed he was hospitalized two years ago "for his stomach." Patient claimed that he has no known allergies and no problems with his eyes, ears, nose, neck, heart, lungs, stomach, intestines, or urinary system. He claimed that he does have a "nerve disorder." "I break out on the arm with a rash when I get nervous." Patient claimed that he is on no medication at present.

8. *PSYCHIATRIC EXAMINATION*: Patient appeared in consultation with neat appearance, dressed appropriately. His cooperativeness was fair. His general manner was one of some preoccupation. Patient's stream of mental activity was accessible. His speech was underproductive. His thought was logical, relevant, and coherent. Emotionally, patient appeared to be depressed and somewhat apathetic. His affect was appropriate. The content of the patient's thought showed evidence of obsessions. His sensorium was normal with clear consciousness. Orientation was normal to time, place, and person. Memory was normal. Intellectually, patient was able to read, write, and calculate to his level of education. Intelligence appeared to be average. The patient's judgment was developing. Patient worries about, "My in-laws, the things I do wrong, and trying to repair them." He is angry when, "People make me mad by the things they do or something I can't do and I know I can." He dislikes, "Little things my wife does or my mother or brother run over me; they take too much and give nothing." He is sad about, "Myself, I feel too sorry for myself at times." When patient was given three wishes, he wished two: "(1) Children; (2) A spot of life." He claimed that he has some dreams at night of, "My dad after he died." When asked why he thought he was seeking professional help, he claimed, "I have tried many other things." When asked for some of the explanations of his problems, he claimed, "By trying not to do them but it doesn't work." He added, "The bar-life is bad for me. But I like it in a way."

8A. *NEUROLOGICAL EXAMINATION*: Cerebral, as described above. Cerebellar, negative. Gait, normal. Cranial nerves: Visual acuity, within normal limits. Acoustic acuity within normal limits. Patient smiled symmetrically. Sensory: Grossly intact to pain, touch, pressure, and position. Motor: No evidence of atrophy, hypertrophy, spasm, or tics of muscle groups of neck, arms, trunk, or legs. Reflexes: all physiological.

9. *PSYCHOLOGICAL TESTS*: DIM = 43, indicating slight depressions; MMS = 50, indicating mild to moderate anxiety; OIT = 85, normal form perception.

10. *ELECTROENCEPHALOGRAM*: Given on 6-19-60. This record is atypical with continuous alpha stereotypy in a diffuse fashion for the entire brain of either cortical or sub-cortical origin. (40021.09). *Clinical correlation*: (1) This is atypical brain functioning. (2) The findings are non-specific. (3) This type of record is seen in lowered vigilance prone to give "neurotic symptons," but not specifically diagnostic and must be correlated clinically. *Recommendations*: (1) Neuropsychiatric observation; (2) Repeat EEG is clinically indicated.

11. *DIFFERENTIAL DIAGNOSIS*: Personality disorder, episodic excessive drinking (APA code 303.0).

12. *TENTATIVE PROGNOSIS*: Guarded.

13. *INITIAL INTERPRETATION*: Al is the product of a very underprivileged home, with a brutal and alcoholic father and an equally confused alcoholic mother. He was raised in chronic depression and has acquired a personality pattern not too dissimilar to the parents. In spite of his psychological handicaps, he had good enough sense and good fortune to marry a rather reasonable woman who is both patient and concerned, in spite of the fact she had been told to get a divorce because of Al's behavior of not facing the family responsibility and episodic excessive drinking. Al's life situation represents a typical case of personality disorder with alcoholic episodes and marital quarreling and potential divorce, which forced him at present to seek professional help.

14. *RECOMMENDATIONS*: (1) Antabuse; (2) Marital therapy; (3) Group therapy; (4) Telefusion couch therapy.

15. *REMARKS AND PROGRESS NOTE*: The patient was given couch therapy to bring forth some of the traumatic experiences in his life which led to the formation of his personality disorder.

CASE 2

NAME: L. A.
AGE: 30
SEX: Male
RACE: Caucasian
MARITAL STATUS: Married
EDUCATION: High school
OCCUPATION: Salesman
BIRTHPLACE: Michigan
IDEOLOGY: Christianity,
REFERRED BY: Hospital
INFORMANT: Patient, reliable, willing

1. *CHIEF COMPLAINT*: (Given on 8-23-68). "Drinking problems and violence."

2. *PRESENT DISTURBANCE*: According to the patient, "I have been drinking for some years, starting in the navy. I caught my first wife with another man and I could have killed them both. I was drunk and I got drunk more often after that. I get drunk every three to six months and then I even call up the police and threaten to kill them or my family." Patient's wife stated, "Larry also has tried to kill himself. When he is sober he is a good father and a good husband. We are having financial problems because he drank up all of our money." Apparently patient has been treated for alcoholism at local hospitals but has not responded in any meaningful way to his therapy.

3. *PAST DISTURBANCES:* Patient claimed that he has had the same problems before. While in the navy, patient claimed he had been placed in the stockade on several occasions because of threats against the military police.

4. *PREVIOUS TREATMENT*: Patient claimed he has had counseling and that he has been to a mental health clinic and to private psychiatrists. When asked why he did not continue working with these doctors, patient claimed, "I didn't feel I needed help. There were financial reasons, and I didn't have time." Patient claimed he is taking "nerve pills" and tranquilizers; "I take a pill once a day for my drinking and one three times a day for my nerves."

5. *PERSONAL HISTORY*: Patient was normally conceived. His birth was normal, full term, with no birth complications or birth injury. He attended nursery school at the age of five; adjustment was "normal." He began grade school at the age of six and adjustment was "normal." He completed through the eleventh grade in high school. Patient began dating at the age of sixteen and steady dating at the age of seventeen. His first important emotional relationship was at the age of seventeen, lasting one year. When asked how it ended, he claimed, "We broke up." He was married at age twenty in a church with approval of both sides of the family. He was divorced in 1960 "through a French court." His first job, outside the military, was at age twenty-eight, as a "factory worker." His last job was at age thirty as a truck driver. He was brought up as a Protestant. He is now practicing as a Catholic. His political views are "none." His outlook on life is "a very dim view at this time." His self-description is, "I am awful tense inside at this time, I don't know why."

6. *FAMILY HISTORY:* Father, age sixty-one, deceased. Stepfather, age forty-nine, grade school education, married, life adjustment "poor; in process of divorce." Mother, age forty-seven, two years of college, life adjustment "poor." Brother, married, life adjustment "good." Sister, age thirty-five, married, life adjustment "fair." Brother, age forty, married, life adjustment "good." Wife, age twenty-five, two years of college, life adjustment "fair." Sons, age five, four, and eighteen months.

7. *SYSTEMIC REVIEW*: Patient has a fair complexion and medium body build. He is 5′10″ tall and weighs 180 pounds. He has brown hair and blue eyes. When asked if he used alcoholic beverages, he claimed

"excessive at times." He claimed that he has had treatment for a liquor habit. His previous hospitalization was "in a mental ward for drinking." He claimed that he has no known allergies. He has no problem with his eyes, ears, nose, neck, heart, lungs, stomach, intestines, or urinary system.

8. *PSYCHIATRIC EXAMINATION*: Patient appeared in consultation with neat appearance, dressed appropriately. His general manner was one of preoccupation. His general activity showed some hyperactivity. Patient's stream of mental activity was accessible. His speech was normal. His thought was logical, relevant, and coherent. Emotionally, patient appeared to be depressed and anxious. The content of the patient's thought showed evidence of obsessions, "old girl" and parents. His sensorium was normal with clear consciousness. Orientation was normal to time, place, and person. Memory was normal. Intellectually, patient was able to read, write, and calculate to his level of education. Intelligence appeared to be average. The patient's judgment was good, and his insight was quite perceptive. Patient worries about "home and finances." He fears "losing my family." He is angry or gets angry about "being used." He claimed that he has no strong dislikes. He is jealous of his "wife." He is envious of his "brothers." He is sad at "wanting to be someone and unable to." When patient was given three wishes, he wished: "1. To be someone; 2. to have a good education; 3. to have a good job." When asked why he chose these wishes, he claimed, "It's what I want most in life." Patient claimed that he does have dreams at night, "mostly about the navy service and my work." He claimed that he has repetitive dreams "about navy service." When asked why he thought he was seeking professional help, he claimed, "because I need help." When asked for some explanations of his problems, patient claimed, "unknown."

8A. *NEUROLOGICAL EXAMINATION*: Cerebral, as described above. Cerebellar: negative, gait normal. Cranial nerves: visual acuity within normal limits. Patient smiled symmetrically. Sensory: grossly intact to pain and touch and position. Motor: no evidence of atrophy, hypertrophy, spasm, or tics of muscle groups of neck, arms, trunk, or legs. Reflexes: all physiological.

9. *PSYCHOLOGICAL TESTS:* MMS = 57.5, indicating moderate anxiety. DIM = 54, indicating mild to moderate depression. OIT = 70,

10. *ELECTROENCEPHALOGRAM:* Given on 10-28-60. This record is atypical, slightly dysrhythmic increased low voltage fast activity and abrupt beta-alpha transitions in a diffuse fashion involving the entire brain with possible sub-cortical sign (43131.00). *Clinical correlation*: This is atypical brain functioning. 2. The findings are not specific. 3. There is no gross evidence of organic brain damage or structural lesion. 4. However, minimum brain dysfunction cannot be ruled out, especially in the presence of medication. 5. No convincing evidence of alcoholic encephalopathy. *Recommendations*: 1. Neuropsychiatric exploration. 2. Serial EEG comparison as clinically indicated.

11. *DIFFERENTIAL DIAGNOSIS*: Personality disorder with secondary alcoholism (APA code 303.1).

12. *TENTATIVE PROGNOSIS*: PPS = 56, indicating prognosis is guarded. Personality change would be difficult if wife would not cooperate consistently with personality transformation.

13. *INITIAL INTERPRETATION*: Larry is a product of a very disturbed home. His parents were divorced. He was a loner, and he had tremendous jealousy toward his brother. The early childhood patterning was also mixed up with the marital disturbances of his mother's second marriage. It is most probable he suffered from a severe childhood neurosis, and later it was transformed into a character disorder with both violent tendencies and rage reactions, which he controlled poorly with alcohol. In fact, alcohol intensified his homicidal impulses after each drinking episode.

His personality disorder was partially masked by his life in the navy until his marriage to a woman who duplicated his early childhood life, namely, a confusion of infidelities, divorce, physical violence, etc.

We have here a potentially criminal personality in common parlance, and the probability of transforming such a personality to a socially compatible one is difficult, as most penologists know.

His second marriage was further extension of his disordered personality, with intensification of unresolved family programs. And his current homicidal acting-out is but one of the acute episodes of a chronic anti-social personality, with habitual alcoholism as a compensatory mechanism against homicide and suicide.

14. *RECOMMENDATIONS*: (1) Hospitalization; (2) Drug therapy; (3) Antabuse therapy; (4) Family therapy; (5) Group therapy; (6) Telefusion couch therapy.

15. *REMARKS AND PROGRESS NOTE*: Larry continued in treatment for fifteen months, on a weekly basis, and responded well.

CASE 3

NAME: L. N.
AGE: 34
SEX: Male
RACE: Caucasian
MARITAL STATUS: Married
EDUCATION: 11th grade
OCCUPATION: Tailor
BIRTHPLACE: Michigan
IDEOLOGY: Christianity,
REFERRED BY: Friend
INFORMANT: Patient; patient's wife.

1. *CHIEF COMPLAINT*: "Very depressed and nervous stomach."

2. *PRESENT DISTURBANCE*: Patient's wife claimed, "Len has been depressed and upset for quite awhile. His stomach is upset all the time. It seems as if something is triggering this depression but I can't put my finger on it. He is so depressed it worries me. He says he can't live this way. He is ashamed of going to a doctor for this."

3. *PAST PROBLEMS*: According to the patient, "I have not had similar problems before. I did take the problem of my stomach to other doctors, and they said it was nerves. Medication didn't help me, and I was depressed. That's why I didn't continue with those doctors. I have to get my stomach fixed up."

4. *PREVIOUS TREATMENT*: Patient claimed he has had no previous treatment, except for taking "nerve pills."

5. *PERSONAL HISTORY*: Patient claimed, "My conception was normal, wanted, and out of wedlock. Birth was normal, not a hospital delivery, and no complications. I began grade school at the age of six, adjustment was OK. I dropped out of school in the 11th grade. I began dating at age eighteen with my wife. We were married eight months later when I was nineteen, in a good wedding. We didn't have the approval of both families, because they thought we were too young. My first job was at age eighteen in a factory. My first important job was at age thirty-three, as a tailor. My outlook on life is OK. I am mixed up."

6. *FAMILY HISTORY:* Patient claimed, "My father is sixty years old. He has a fifth grade education, is married, and living in a city near here. His life adjustment is good. My mother is fifty-nine, seventh grade education, good life adjustment. She was a good mother. I have a brother, forty, ninth grade education, married, good life adjustment, and two sisters; one is thirty-six and one is thirty, both are married with good life adjustment. They both have a ninth grade education. My wife is thirty-one, she has an eleventh grade education. We have three children, two daughters and a son, ages ten to fifteen."

7. *SYSTEMIC REVIEW*: Patient is 5′10″ tall, brown hair, brown eyes. He claims to be allergic to seafood. He denies other medical problems. He claims to use alcoholic beverages "very little." Patient claims, "My brother had TB and my mother had diabetes. My present weight is not stationary. I lost nine pounds in four weeks."

8. *PSYCHIATRIC EXAMINATION*: Patient appeared in consultation with neat appearance, dressed appropriately. His general manner was one of preoccupation. His general activity showed somewhat slowed retardation. Patient's stream of mental activity was accessible. His speech was underproductive. His thought was logical. Emotionally, patient appeared to be depressed and anxious. The content of the patient's thought showed evidence of obsessions and psychosomatic complaints. His sensorium was normal with clear consciousness. Orientation was normal to time, place, and person. Memory was normal. Intellectually, patient was able to read, write, and calculate to his level of education. Intelligence appeared to be average. The patient's judgment was defective and his insight was obtuse. Patient worries about "life." He fears "being weak." Patient claimed that he does get angry but did not explain.

He also claimed that he is jealous and envious but did not explain. When patient was given three wishes, he wished for: "1. Health; 2. good marriage; 3. happiness." He claimed that he does have dreams at night "sometimes." When asked to describe one, he claimed, "about being closed in; I don't know." When asked why he was seeking professional help, patient claimed, "I can't seem to find help anywhere." When asked in what ways he has tried to solve his problems, he claimed, "By church and doctors."

8A. *NEUROLOGICAL EXAMINATION*: Cerebral, as above. Cerebellar, sensory, motor and reflexes, negative.

9. *PSYCHOLOGICAL TESTS*: DIM = 45, indicating mild depression: MSO # 42.5 indicating average reality testing; MMS # 55, indicating mild to moderate anxiety; OIT # 86, normal form perception.

10. *ELECTROENCEPHALOGRAM*: Given on 5-26-60. This record is atypical with alpha stereotypy over alpha background (40020.00). *Clinical correlation*: (1) This is atypical brain functioning. (2) The findings are non-specific. (3) There is no evidence of structural focal-brain damage. (4) The atypical alpha stereotypy has been associated with neuropsychiatric disturbances, but not specifically diagnositc. (5) This record, of course, cannot rule out post-traumatic effect. *Recommendations*: (1) Your clinical management. (2) Repeat EEG should there occur objective neurological signs, or otherwise clinically indicated.

11. *DIFFERENTIAL DIAGNOSIS*: Depressive neurosis (APA code 300.4).

12. *TENTATIVE PROGNOSIS*: Good with treatment.

13. *INITIAL INTERPRETATION*: Len is a bright, normally intelligent person with tremendous feeling of insecurity and inferiority, born partly of Afro-American background and raised with the psychology of physical force as the best way of solving problems. This was true to such an extent that his father encouraged him to become a champion boxer. He was unsuited to be a boxer because he had frequent attacks of anxiety and feelings of a lack of true identity until he changed his profession to become a tailor. However, this came into conflict with his value system. In his case, the two personality structures are quite clearly demarcated: Len, the boxer; Lou, the tailor. This kind of sensitivity and this kind of aggressiveness are mutually contradictory, so while he is doing his work as a tailor he has surges of aggression towards his work environment, including the people involved, toward whom he has developed feelings of inferiority in order to control his aggressive impulses. Because of this personality crisis and identity disturbance, he developed a psychoneurosis which can be best described as a depressive type. He is seeking therapy only to relieve his symptoms of depression but has very little insight regarding personality modification. In summary, then, we have here a bright, normal, intelligent individual who has been caught between transitional cultural changes, the value systems from the brave warriors of Afro-American heritage emerging as the boxer, to the well-controlled and appearance conscious way of life in tailoring. The depression is well-measured on the MMPI (D scale of 84),

which gives us a full summary of his current problems.

14. *RECOMMENDATIONS*: (1) marital therapy, involving his wife; (2) couch therapy to assess his early personality pattern; (3) group and family videotherapy to aid personality transformation so that he could succeed in becoming in complete harmony as a tailor; (4) chemotherapy, a course of anti-anxiety and anti-depressants would be of help in accelerating therapeutic progress; (5) Telefusion couch therapy.

15. *REMARKS AND PROGRESS NOTE*: Patient had a course of couch therapy and concurrent marital therapy with good results. All his therapies were videotaped and videolinked.

CASE 4

NAME: B. E.
AGE: 36
SEX: Female
RACE: Caucasian
MARITAL STATUS: Married
EDUCATION: 10th grade
OCCUPATION: Housewife
BIRTHPLACE: Michigan
IDEOLOGY: Christianity,
REFERRED BY: Family physician
INFORMANT: Patient

1. *CHIEF COMPLAINT*: (Given on 3-2-60). "I am picking on my husband and destroying my second marriage. I don't know why."

2. *PRESENT DISTURBANCE*: Patient claimed, "I don't know why I am picking on my husband because he is a good man, he doesn't deserve it, yet I can't help it. Every morning I'm looking for something to pick on him for. He gets upset. He treats my children, all three of them, like they were his own. I really like my marriage. I don't want anything to happen to it."

3. *PAST PROBLEMS*: According to the patient, her mother left when patient was twelve years old. "She just packed her bags and walked out. I was left with my father and four brothers and sisters. I also took care of the house. I haven't talked with my mother very much, only recently." Patient also claimed, "My first marriage failed. I don't know why. I got involved with this man when I was seventeen and he was seven years older. He was involved with the law. I thought anyone was entitled to a mistake. I went with him. The marriage never worked. He had to go overseas. Finally I decided I had had enough and we were divorced four years ago. My present husband and I had known each other about ten years. We had double-dated but never dated. We were married last year."

4. *PREVIOUS TREATMENT*: Patient claimed she has not been to another psychiatrist but has seen doctors for "my knee. They thought I

had arthritis. They thought I should have a hysterectomy. I didn't do it. They put me on nerve pills and pain pills. They asked me if I would like to see a psychiatrist because I had blow-ups like biting my fingernails and yelling and screaming. I have had Thorazine twice a day for nerves."

5. *PERSONAL HISTORY*: Patient claimed her conception was "normal, wanted and in wedlock. Birth was normal, full term and a hospital delivery, with no complications. I began grade school at age four and completed through 10th grade in high school. I began dating at fifteen and my first important emotional relationship was at seventeen with my first husband. We were married when I was seventeen, by a judge. My family did not approve because my husband was an alien. My first job was at thirteen as a babysitter. My first important job was at twenty-one as a factory worker. I was brought up as a Protestant. My outlook on life is very dull at times and at other times very good. I am hard to get along with at times and nothing pleases me."

6. *FAMILY HISTORY*: Patient claimed, "My father is fifty-three, unemployed with an 8th grade education; he is divorced. His life adjustment is poor. My foster father is sixty-three, a construction worker with a good life adjustment. My foster mother is fifty, a bookkeeper, remarried, with a fair life adjustment. I have a sister, thirty, married with a poor life adjustment; brother, twenty-eight, married, fair life adjustment; brother, twenty-five, married, mechanic, fair life adjustment; sister, twenty-two, college student, remarried, fair life adjustment. My husband is twenty-seven, 10th grade education, first marriage, fair life adjustment. I have two sons by my first marriage, eight and five years old, and a daughter, age four."

7. *SYSTEMIC REVIEW*: Patient claimed she has a fair complexion, good body build, weight 146 pounds, height 5′7″, brown hair and green eyes. Her last medical examination was five months ago. She has had a tonsillectomy. She wears glasses for nearsightedness. She has no problem with ears, nose, neck, heart, lungs, stomach, intestines, or urinary system. She does have headaches. Her weight has been stationary for ten months. She began menstruation at twelve and her last period was three months ago. She is not pregnant at this time. Her menstrual period lasts six to eight days. She has had three pregnancies.

8. *PSYCHIATRIC EXAMINATION*: Patient appeared in consultation with neat appearance, dressed appropriately. Her cooperativeness was good. Her general manner was one of seeking help with sincerity. Her general activity showed to be within normal limits. Patient's stream of mental activity was accessible. Her speech was productive. Her thought was logical, relevant, and coherent. Emotionally, patient appeared to be depressed, anxious, and irritable. The content of the patient's thought showed evidence of obsessions with "first husband, violent father, and a broken home." Her sensorium was normal with clear consciousness. Orientation was normal to time, place, and person. Memory was normal. Intellectually, patient was able to read, write, and calculate to her level of education. Intelligence appeared to be average. The patient's judgment

was fair and her insight was developing. Patient claims she worries about "bills and my family. I fear that I will destroy my marriage. At times, everything makes me angry. At times I dislike everything. I hate my ex-husband. Nothing makes me envious. I am unhappy and sad when I worry all of the time." When patient was asked what her three wishes would be, she said, "1. To be a better wife; 2. to be a better mother; 3. to stop worrying all of the time." She claimed she wished these because "I want to better myself." She dreams "that my father is sick. I repeatedly dream that my father is alone and sick." Patient claimed she is seeking professional help "to learn to understand myself and save my marriage. I pick at my husband all the time and am a very unhappy person at times. I have tried to stop picking and enjoy my husband and children. I am very hard to please but I know my husband is doing his best. I don't seem to be doing mine."

8A. *NEUROLOGICAL EXAMINATION*: Cerebral, as described previously. Cerebellar: Negative. Gait: normal. Cranial nerves: Visual acuity, within normal limits. Acoustic acuity within normal limits. Patient smiled symmetrically. Sensory: Grossly intact to pain, touch, pressure, and position. Motor: No evidence of atrophy, hypertrophy, spasm, or tics of muscle groups of neck, arms, trunk, or legs. Reflexes: all physiological.

9. *PSYCHOLOGICAL TESTS*: MMS = 70, indicating severe anxiety; DIM = 58, moderate depression; MSQ = 67.75, poor reality perception; PPS = 72, good prognosis; OIT = 13, indicating poor perception.

10. *ELECTROENCEPHALOGRAM*: Given on 3-4-60. This record is abnormal with theta showing, consistently in a diffuse fashion for the entire brain, or either cortical and/or subcortical origin (79201.09). *Clinical correlation*: (1) This is mildly abnormal brain functioning supportive of diffuse encephalopathy. (2) This is supportive of chronic brain syndrome, including impairment of memory, judgment, orientation, etc., but must be correlated clinically. (3) There is no evidence of focal disturbance. *Recommendations*: Neuropsychiatric exploration, especially regarding early organic brain syndrome. Repeat EEG as clinically indicated.

11. *DIFFERENTIAL DIAGNOSIS*: (1) Schizo-affective schizophrenia (APA code 295.74). (2) R/O Schizoid personality (APA code 301.2). (3) R/O Depression neurosis (APA code 300.4).

12. *TENTATIVE PROGNOSIS*: Poor, without treatment, in view of chronic history and gross disturbance of family relationship and disturbed marital relationship and divorce and subsequent confusion of identity for herself, her husband, and her children.

13. *INITIAL INTERPRETATION*: Bethany is a woman of average intelligence, born in a very deprived home with unemployed and divorced father and mother who was ineffectual. The foster father gave reasonable support, but family instability has been a pattern, including her siblings. Her first marriage was disastrous, and as a result her stability came under tremendous pressure and showed obvious symptoms of depression and personality disorganization. It is conceivable that she has always had a schizoid personality, and with adverse events in her family

life, she developed depression. However, her OIT and her MMPI both indicated possibility of perceptual disorder, including early psychotic reaction. She could be regarded best as schizo-affective schizophrenia, depressed type.

14. *RECOMMENDATIONS*: (1) Marital therapy with psychosynthesis.

6. PSYCHOSYNTHESIS: Psychoanalysis with Telefusion

It seems so very natural that psychosynthesis should develop out of psychoanalysis. Today, psychosynthesis is arriving on the scene, as if it were to rescue psychoanalysis from its often-pronounced demise by such self-appointed ideologists as Eysenck, Szasz, Wolpe, and others. They would like to continue to offer opposition to Freud and lead a public hue and cry against psychoanalysis.

However, the "public" has ridiculed psychoanalysis, and made fun of the first psychoanalyst since the days of Vienna. Such critics, professional or lay, have missed the central point that psychoanalysis has opened up the diverse approaches to psychotherapy, the rational approaches to human irrationalities, to the rich data collection of personality development, etc. and now to psychosynthesis[3].

PSYCHOSYNTHESIS

What is psychosynthesis? Psychosynthesis includes psychoanalysis but goes beyond it. Psychosynthesis, unlike psychoanalysis, looks for higher personality synthesis and construction. Psychoanalysis explores and studies in depth the evolution of a personality and its development, whereas psychosynthesis studies the possibilities and potentialities of the higher transformation, higher synthesis of a personality.

Psychosynthesis is also behavior therapy, but goes beyond behavior therapy. Behavior therapy aims at the modification of the overt behavior of the personality without analyzing in depth the evolution and development of the personality.

Psychosynthesis goes beyond behavior therapy, for it is not only an attempt to determine and redirect the behavior pattern of a personality, but it also changes the program within the brain. Behavior therapy is aimed at the adjustment in the psychosocial environment of a person, whereas psychosynthesis aims at transforming not only the self to a higher level of integration, but also requiring fundamental changes in the direction of reconstruction of the social relationships with significant others in harmony with his new personality structure.

Psychosynthesis is chemotherapy, but it goes beyond chemotherapy. Chemotherapy attempts to modify, strengthen, stabilize, or to normalize any disturbed functioning of the central nervous system, whether in cortical, subcortical, hypothalamic or the interconnections of the subsystems. Psychosynthesis goes beyond. Psychosynthesis aims at the stabilization and normalization of the central nervous system as a highly integrative system for the purpose of information processing, goal orientation and goal direction for a higher social integration for evolution.

Psychosynthesis is psychotherapy but goes beyond psychotherapy. Psychotherapy aims at the treatment of illnesses, whereas psychosynthesis aims at transformation of personality to any possible level of achievement and to any horizon that the individual has the capacity to attain. Psychotherapy aims at the removal of disorders and illnesses, diseases or complexes, whereas psychosynthesis aims at integration, transformation and organization of the individual as a creative force with a new identity.

Psychosynthesis is group therapy, but goes beyond it. Group therapy aims at personality modification through group dynamics and socializing pressures. Psychosynthesis is social psychiatry but goes beyond it. Social psychiatry aims at community participation and group relationship. However, psychosynthesis provides the theory and unites the eclectic techniques for achieving personality transformation. In personality transformation, an individual takes on a new name and a new identity by psychosynthesizing the contradictory aspects of a more primitive personality.

In short, psychosynthesis is the cybernetic union of all eclectic elements of psychiatry, fusing psychoanalysis, psychotherapies, family therapy, chemotherapy, electrotherapy, behavior therapy, group therapy and community mental health into a television feedback system[5]. This is called the television system of psychosynthesis. The reader at this point might feel the need for clarification. What theoretical framework can include such diverse and seemingly contradictory approaches? The mass-energy equivalence equation, $E = MC^2$, of Einstein. I have discussed this formulation of psychosynthesis elsewhere[4].

Psychosynthesis is the attempt to look upon all psychological (psychoanalysis, psychotherapy, group dynamics, etc.) approaches as the modification of patterns on the energy (E) side of the Einsteinian equation, while the organic approaches (chemotherapy, electrotherapy, etc.) are trying to influence mass (MC^2) of the same equation. The psychosynthesis model views personality transformation essentially as energy pattern transformation. Let us look at the nature of the psychosynthesis model.

THE PSYCHOSYNTHESIS MODEL

Psychosynthesis will bring forth the contributions of psychoanalysis with true aim, through its unified theory on mind and behavior. The theory of psychosynthesis brings into focus three basic energy functions as follows:

$$E_i(E_s) \rightarrow E_o.$$

E_i represents the input energy pattern, transmissible into the system; E_s is the stored energy pattern of genetic and experential memory; and E_o is the output energy pattern, i.e., behavior.

The model of psychosynthesis is based on the cybernetic principle of continuous feedback and modification.

Personality is defined as a time-series of E_s with consciousness as a goal-directed program for self-creation, normal or otherwise. In psychosynthesis the personality is viewed as an energy matrix of on-and-off electrical discharges projected by the neurons of a recycling pattern of discharges. This matrix is a matrix of many dimensions in terms of sensations, perceptions, conceptions with linguistic, emotional, behavioral and motor multiplex integration. This unified energy model resolves the conflicts as to the means which can modify a given personality. It can be modified verbally, chemically, electrically, or by any combinations thereof, by modifying E_i.

This explains why a personality may be modified by analysis, suggestion, hypnosis, group therapy, chemotherapy, electrotherapy, or community experiences, since E_s of any person may be modified by any means, and the results and efficacy depend on many variables. And one of the important variables is technology.

TECHNOLOGY OF PSYCHOSYNTHESIS

Psychosynthesis cannot go far without technology. This is seen clearly with the classical psychosynthesis of Assagioli(1), whose work is completely overshadowed by psychoanalysis and other verbal psychotherapies. Modern psychosynthesis or cybernetic psychosynthesis(5) points to the fact that memory (E_s) of both genetic information of evolution and experiential information of the individual requires modification, if basic personality modification is to be achieved. Cybernetic psychosynthesis brings out the need of loosening and erasure of consciousness and memory as important elements of influencing the modification of E_s, hence the energy pattern of a personality. Erasing and loosening of consciousness and memory could not be achieved without the advancement of technology.

Psychosynthesis couldn't be achieved in the early days of psychoanalysis, because neither electrotherapy nor chemotherapy was sufficiently advanced for such a synthesis. In 1886, Freud stated that "I have placed reliance on physical therapy and found myself helpless in the face of the disappointments that I had with W. Erb's electrotherapy, so rich in the advice and indications." And furthermore, "The treatment by suggestion in deep hypnosis seems to offer at that time sufficient compensation for loss of confidence in electrical therapy."

Freud was unable to synthesize electrotherapy with hypnotherapy at that juncture of technology in the early 1900's. However, technology has changed with the advent of electrical engineering, information theory and cybernetics. The new synthesis of physical therapy (like electrotherapy) and psychotherapy (like hypnotherapy) and depth exploration (like psychoanalysis) has been in the air. Psychosynthesis encompasses physical therapy, psychotherapy and psychoanalysis as a unified system of energy transformation of patterns. Technology has enabled us to achieve psychosynthesis. I agree with anthropologist White[6] that new advances in theory and technique of science depend on synthesis of culturally available elements. The science of psychosynthesis certainly supports this culturological dictum. Cybernetic psychosynthesis requires these key ingredients:

Modern psychopharmacology
Freudian psychoanalysis
Pavlovian conditioning
Electrotherapy of Cerletti
Einstein's mass-energy equivalence
and Shannon's information theory
together with modern television technology.

Technology alone, of course, cannot create synthesis. Cybernetic psychosynthesis is a result of social use and the continuous feedback approximation with individual mental health, marital stability, family stability, and community mental health as the ultimate values. The nature of psychosynthesis requires careful linkage of the subsystems described. The techniques of psychosynthesis are just as important as its theoretical formulation.

TECHNIQUES OF PSYCHOSYNTHESIS

The techniques of psychosynthesis can be summarized as follows:

0. *Drug Therapy*. To stabilize brain function.

1. *Interview Psychotherapy*. This is simply classical psychiatric technique to establish rapport to obtain chief complaints and to relate to the patient for the purpose of evaluating the patient's life and life style and uncovering the nature of causes of his disturbances.

2. *Couch Analysis*. This is the classical Freudian approach in which a patient brings more easily disturbed information programmed in childhood, which may be involved in disturbed psychosexual development or abnormal interpersonal relationships or deprivations due to death of father, mother or other traumatic events in the individual's life. Oedipus complex is only one of such abnormal programs with possible unresolved contradictions for the individual. Classical couch analysis provides information for future modification or erasure.

3. *Family Therapy*. This is a conjoint marital therapy in which information can be explored and confirmed, modified or integrated in relation to the patient's interview or couch analysis.

4. *Electrolytic Therapy*. This is the use of the electrotherapy of Cerletti for the purpose of activating abnormal information obtained either in interview psychotherapy (couch analysis) or family therapy for the purpose of loosening and erasure. Consciousness and abnormal memories are activated just before the application of electrical current.

5. *Postlytic Family Therapy*. This is the first step of union of electrotherapy, psychotherapy, family therapy and classical conditioning in one session. In the postlytic session, the patient is bottle-fed like an infant for the purpose of establishing a new and significant relationship with a catalyzer or programmer, who may be a husband or wife or original parents.

6. *Transnomation*. This is a therapeutic name-change, usually the first name. For example, Nancy's name can be changed to Carol, who then becomes the new personality with a new name.

7. *Group Therapy for Psychosynthesis*. This is group therapy composed of families who have gone through individual therapy with at least one member of the family; sometimes both. In this group situation, the person with the new name is so introduced to establish and strengthen the new identity in a new situation.

8. *Mental Health Mini-Community*. This is the fusion of three psychosynthesis groups into a mini-community to promote community mental health. They may establish a coffee shop by using their financial resources to form a corporation. The families use the coffee shop to meet and to amplify and to mutually assist the families, not only with employment and social activities, but also personal contact.

The above techniques are incorporated into psychosynthesis by the use of television, so that every family is known to every other family through the closed-circuit television which serves as feedback correction, amplification, storage and intensification of the therapeutic effects of any of the above.

CONCLUSION

The nature of psychosynthesis reflects the process of evolution through selection, integration, extension of socially and therapeutically useful techniques with goal direction of creating community mental health.

The nature of psychosynthesis may be seen as continuous evolution of life and of evolving homogeneous, isolated individuals into a complex, community life through information processing, feedback, and growth with clear goal direction.

The nature of psychosynthesis is determined by the psychiatric advances in psychoanalysis, psychotherapy, electrotherapy, chemotherapy, group therapy, and community mental health principles linked by television and technology.

The nature of psychosynthesis brings out one important ingredient that is a departure from all previous psychosynthesis, including psychoanalysis and classical psychosynthesis, namely, aiming at modification of personality, by not only adding information into the patient's or person's mind, but also the subtraction, the loosening, and erasure of information from the patient's mind. This points out the important direction of research in electrotherapy, chemotherapy, or combined approaches to seek better ways of loosening and erasing abnormal information from the brain.

The nature of psychosynthesis also brings out a new philosophical dictum in personality theory, namely, "memini, ergo sum.": I remember, therefore, I am. It is one thing to have discovered the Oedipus complex, but it is quite another to erase it. The nature of psychosynthesis advocates not only the interpretation and analysis of abnormal personality complexes, but the point is how to change them.

SUMMARY

Psychosynthesis is the telefusion (television-linked cybernetic system) of many theoretical and technical elements, such as the psychoanalysis of Freud, the behavior therapy of Pavlov, the electrotherapy of Cerletti, the drug therapy of the Space Age in the light of cybernetics of Wiener, and information theory of Shannon as a unified system of brain-mind equivalence in the same sense of mass-energy equation of Albert Einstein, with life as our ultimate value.

However, in the dialectics of nature, along the line of thought of synthesis, the laws of physics and relativity should have applicability to psychology and psychiatry. The laws governing matter should also govern the mind, and the laws of the mind shall control matter. By analogy, the relationship and magnitude of work and energy in psychoanalysis to psychosynthesis lies in their dialectic relationship as the power of fission is to the power of fusion in atomic physics. The source of synthesis or fusion energy is released by analysis of fission; and this fusion energy was wished for and anticipated by Assagioli in classical synthesis, but unrealized until the development of videotechniques and telefusion technology created by the electrical and electronic workers and mental health workers, videotechnicians, and videotherapist-psychiatrists.

The power of telefusion is great, as it has been applied to amplify couch analysis to telefusion couch therapy, which may be used on a massive scale in service of the workers and all applicants and patients who need depth psychotherapy to combat depression, anxieties, alcoholism, drug abuse, divorce, boredom, etc., of the industrialized societies and industrializing world under the shadow of the thermonuclear bomb.

The next chapter will describe the videotechniques of *telefusion couch therapy* in detail.

REFERENCES

1. R. Assagioli, *Psychosynthesis* (New York: Hobbs, Dorman & Company, Inc., 1965), p. 194.
2. H. C. Tien, "Pattern Recognition and Psychosynthesis." *Am. J. Psychother.* 23: pp. 53–66.
3. H. C. Tien, "Psychosynthesis: A Cybernetic System of Psychotherapy" (Sixth Annual Rocky Mountain Bioengineering Symposium, Laramie, Wyoming, 1969), pp. 78–81.
4. H. C. Tien, "Psychosynthesis: Television-linked Cybernetic System of Psychotherapy" (Proceedings of International Congress of Cybernetics, London, England, September 1969).
5. H. C. Tien, "Psychosynthesis: Television-Cybernetic System of Psychotherapy" (Third Annual Symposium, American Society of Cybernetics, Gaithersburg, Maryland, October 1969).
6. Leslie A. White, *The Science of Culture, A Study of Man and Civilization* (New York: Farrar, Straus and Giroux, 1949)

7. TELEFUSION COUCH THERAPY

Telefusion enables an individual to do both analysis and synthesis on himself.

Telefusion couch therapy is an outgrowth of classical couch analysis, but goes beyond it. It is a modern technique, combining analysis with synthesis. Telefusion couch therapy is goal-directed, specifically toward analyzing the undesirable personality with a goal of building a higher personality out of free association and free communication.

There are three types of telefusion couch therapy: (1) Primary telefusion couch therapy; (2) Ordinary telefusion couch therapy; and (3) Booster telefusion couch therapy.

(1) *Primary telefusion couch therapy*: Here both the synthesist (i.e., the therapist) and the synthesand (namely, the patient) are TV-taped and

also TV-linked, with immediate feedback through television. This, of course, enhances the nature of couch therapy, for the patient has even more privacy. Not only are his eyes averted from the therapist, he is actually in a different room. Free association is even more encouraged, but at the same time, a new dimension is added: namely, free communication. The therapist is able to give information to assist the patient to speed up the process with a goal orientation. Primary telefusion couch therapy is defined by its technical aspect, immediate feedback.

(2) *Ordinary telefusion couch therapy*: Here the synthesand has an opportunity for free association and free communication through TV tapes. The patient is TV-taped, but the therapeutic process is being conducted through television by the synthesist on tape. This has no immediate feedback, but it maximizes privacy and the patient is further encouraged for self-analysis with technological help. The feedback is intermediate. By that we mean that at the end of the ordinary telefusion couch therapy session, the therapist and the patient meet to discuss optimally regarding the direction of therapy.

(3) *Booster telefusion couch therapy*: This is similar to ordinary telefusion couch therapy, except there is no intermediate feedback. The feedback may be postponed for a month or three months. Booster telefusion couch therapy is also used in the absence of the therapist when continuity can be achieved. Booster telefusion couch therapy can also be used when a patient needs three or four sessions a week, which can be intensified by supplementing booster or ordinary telefusion sessions to the primary session.

All three telefusion couch therapy techniques are goal-directed and client-structured. We are going to describe here the beginning of a typical series of twelve sessions of telefusion couch therapy, which may last anywhere from one month to three months, depending on whether the session is conducted weekly or three times a week.

PRIMARY TELEFUSION COUCH THERAPY

(Patient on couch sees and hears therapist on television.)

THERAPIST: This is your first telefusion couch session. If you will just follow my comments, you will find this is going to be very helpful and effective for you. Telefusion simply means we are trying to put things together on television. Couch therapy is a classical approach. In telefusion couch therapy, we use a classical position with the added advantage of exact production, by TV, of the information you produce by free association. We use this instead of writing notes, which are really quite inadequate, as we have found. With television, we also have free communication. Later on, I will give you free feedback.

You will be entirely alone in the room. At the same time, you will be able to hear my directions as in classical approach. Would you close your eyes now? The idea is for you to free associate. By that I mean, speak about anything, four-letter words if you want to, hate, anger, the past, present, future. Let your mind be free. We will record everything. Even if it's a jumble, it doesn't make any difference. Even if you feel you don't make sense, that's fine. Would you start now?

(Patient free associates while therapist listens in other room.)

Patient in primary telefusion couch therapy free associates and receives free communication from the therapist, who is in a different room. At the same time, both patient and therapist are being recorded on tape to provide later feedback for the patient.

THERAPIST: You have done very well in this session. I am going to play this tape back to you so you will see very clearly where your personality is, and where you are trying to go. To build a higher personality, you need information, you need correction, you need feedback. That is the purpose of couch analysis.

You have done very well in this session. I am going to play this tape back to you so you will see very clearly where your personality is, and where you are trying to go. To build a higher personality, you need information, you need correction, you need feedback. That is the purpose of couch analysis.

TELEFUSION COUCH FEEDBACK

The patient watches a TV tape of his couch therapy and uses a printed form (see below) to record his impressions:

As illustrated above, in telefusion couch feedback, patient watches TV tape of his earlier telefusion couch therapy session and writes down statements and impressions that have significance for him.

MICHIGAN INSTITUTE OF PSYCHOSYNTHESIS
NOTES ON TELEFUSION COUCH FEEDBACK

NAME________________ DATE________________

INSTRUCTIONS: Please take notes as detailed as you can. Give direct quotations when significant to you. Record dreams, fantasies, conflicts, and recurrent statements or repeated problems. *Use both sides of the page if possible.*

1. Please describe the beginning of the session:

2. What thoughts, feelings and ideas were brought out?

3. The names of people, relatives, friends, teachers, etc., mentioned:

4. Write down any topic or idea that is repeated in the session:

5. Write down unsolved problems, questions:

6. The areas I would like to explore in the future are:

7. The last thoughts on the tape were:

8. (Please circle your comment regarding the feedback):
The feedback itself was: Excellent / Good / Fair / Poor
For me, the feedback was: Very helpful / Helpful / Somewhat helpful / Not helpful

9. I would like to suggest some changes to help my treatment:

(If you need more space, use back of page.)

"1. Seminary experience, having emotional troubles, seminary was a 'shock.' I was always homesick, an emotional child. I longed for home as if it were paradise. I was adjusted to home and *only* to it. I couldn't take the living with other guys, I didn't want to 'give and take' with anybody. I was full of prejudices. As Dad said, 'Never question an order, just do it.' Obedience was everything. I was always loyal to family, church, and party. Morality was anti-sex and anti-pleasure. I knew nothing about sex except it was 'bad and dirty.' I was rebellious inside toward Dad but felt I was wrong to feel that way. I resent all authoritarian figures and rebel against their orders (at least inside). No one is going to push me around. My lower personality is antagonistic, resentful, defensive, belligerent, and pessimistic. My higher personality is optimistic and tries to like people.

"2. My rejection of authority figures; all those who wield power over others.

"3. Dad; Mother; my brother.

"4. Prejudice against others not as upright or puritan as our family. People who left the church were awful per se.

"5. I can't seem to let out my higher self. If someone rubs me the wrong way, I become fearful and angry.

"6. How my father's death was also the partial death of my lower personality and a new life for me.

"7. To try to separate the lower and higher personalities.

"8. For me, the feedback was good, helpful."

DISCUSSION

The above scheme shows how television can revitalize the classical approach of couch therapy. Telefusion couch therapy provides flexibility, continuity, and economy for all patients who are motivated for intensive therapy, when the cost function and the availability of analyst become reality problems.

However, telefusion couch therapy contributes not only in terms of availability and economy and towards intensive psychotherapy, but it also provides further practical and theoretical development in terms of objective analysis and synthesis.

Telefusion couch therapy brings forth sharply the development of psychoanalysis to psychosynthesis. With psychoanalysis, we have free association, and with psychosynthesis, we have free communication. It is the fusion of both analysis and synthesis. By using the modern technology we are able to bring the insights of psychiatry to a greater number of people, and by changing our own orientation significantly, we may effect better therapeutic results for all those who come to us.

Telefusion couch therapy is yet at its early developmental state. As the state of art of television technology evolves, we can see even greater development of this approach. Patient acceptance has been good.

8. FROM THE COUCH TO THE COFFEE SHOP

The couch symbolizes psychoanalysis, whereas the coffee shop symbolizes psychosynthesis. When Freud began to have patients lie on the couch to free associate about their problems and conflicts, he made an important advance in psychiatry. However, psychiatry has traveled quite some distance from the couch therapy of Freud to the coffee shop of psychosynthesis today.

The Couch. Of course, many of us have heard a great deal about the Freudian couch and all the jokes and innuendos concerning this approach. But the couch will stay, because it brings forth a wealth of childhood information and secrets of the individual to the analyst for elucidation, clarification, editing and reinterpretation. The couch makes it easier to obtain information from the individual, because he tends to bring out his fantasies, delusions, dreams, secret ambitions, hateful quarrels, love affairs, life episodes that are beautiful or ugly, weird or ecstatic. The couch naturally has its limitations. For example, the therapeutic interpretations offered by the psychoanalyst may be denied by the patient. Denial slows down therapy. This denial sometimes may be lessened if another family member goes to therapy. And family therapy is often helpful.

Family Therapy. Family therapy does stabilize the therapeutic program, even in simple drug therapy. For example, a young mother's fourth child came as a surprise to her, because she had "forgotten" to take her contraceptive pills. Her husband could have encouraged her to stay on the pills. Of course, family therapy has its own limitations. A family in therapy is still a family in isolation. It is not difficult to see that the child who goes against the values of the parents often uses the supposed values of his friends or neighbors to weaken prematurely the parental control and guidance in an undesirable direction. Group therapy involving several families has a corrective effect on the family in isolation of a highly mobile society.

What we need is family psychiatry. In the professional world of specialization, there is a great tendency for more specialization and diversification into sub-specialties. This is also true in psychiatry. Today we have so many sub-specialties, such as psychoanalysis, child psychiatry, adolescent psychiatry, marital therapy, group therapy, drug therapy, electrolytic therapy and various forms of behavioral therapy. And as a result, families are often quite confused as to what methods are effective and where to seek help.

Family Psychiatry. We need a unified approach. Family psychiatry offers this approach, whether the family problem involves a teenager who is on drugs or a father who is alcoholic or a mother who attempts suicide. These problems are often interlocked. To seek separate agencies for solutions is neither economical nor optimal. The family psychiatrist has to offer a comprehensive service, including interview psychotherapy, couch analysis, drug therapy, marital therapy, family therapy, group therapy, etc. Family psychiatry requires all the eclectic elements of psychiatry to help the families to improve the mental health of the individual, the family members and the community. Furthermore, it should provide treatment and stability for prevention, rather than the old-fashioned approach of seeing a psychiatrist as a last ditch effort, as it were. But who is qualified to set up such a comprehensive approach? The eclectic psychiatrist.

The Eclectic Psychiatrist. The eclectic psychiatrist is a forerunner of the family psychiatrist. The eclectic psychiatrist, being aware of the theoretical inconsistency between organic and other psychological schools of thought, has learned to use all elements of psychiatry for the benefit of the families. The gradual emergence of the eclectic psychiatrist in the world of specialization is our answer to expensive and sometimes needless diversifications into sub-specialties. The families often need and request mental health services in all areas of eclectic psychiatry from time to time, for personal stability, family harmony, and child-rearing in our complex society, confronted with the technological stresses of our modern world.

Steady Service. The idea of family psychiatry on a steady basis becomes clear, when brief psychotherapy or crisis intervention can only temporize with the life problems from childhood to old age. Life is then seen as a series of abnormal crises, when in reality, family psychiatry is needed for all families, to meet the mental health of every family member. Every eclectic psychiatrist can and should offer all the needed services by utilizing totally his professional medical knowledge and the latent abilities of the families to build the psychosynthesis system approach to meet the increased demands for service.

The Psychosynthesis System. Psychosynthesis is based on the normative approach. It is as normal to have steady mental health service for every family member as it is for physical health. The family psychiatrist becomes a psychosynthesist when he fuses all eclectic elements of psychiatry into a continuous dynamic feedback system, in which the families can participate in the operation. The psychosynthesis system is built upon the theoretical foundation of brain-mind equivalence concept in which brain (mass) can be affected by mind (energy) and vice versa. One such family psychiatrist could provide a comprehensive service for a hundred families or more on a steady basis, if the families can be helped to form a mini-mental health community by fusion. How is this fusion achieved? With closed-circuit television.

Television. Closed-circuit television is the logical instrument of psychosynthesis. Television to psychosynthesis is like the x-ray machine to the radiologist and the microscope to the pathologist. Television records diagnostic data and amplifies therapeutic information for flexible feedback, control, and utilization as needed. Television, therefore, links and fuses the eclectic elements of modern psychiatry and diverse schools of thought into a rational unified cybernetic system of psychotherapy. Cogently, television enables the family psychiatrist to become a videosynthesist.

The Videosynthesist. The videosynthesist is a psychosynthesist who can give the best mental health service in all areas described and in the most economical way. With the help of television, the psychosynthesist can easily handle 200 patient office visits per week, thus doubling the therapeutic capacity of an average solo eclectic psychiatrist (assuming he handles a maximum 100 patient visits per week). This is a conservative estimate and does not include the beneficial effects of the on-going mental health minicommunity of fifty families. The families run a cooperative coffee shop for mental health purposes by offering part-time jobs and work experiences with recreational, educational and financial returns.

The Coffee Shop. In the coffee shop, the psychodrama of Moreno has become the drama of everyday life, in which the patients and their family members have actually put their time, money and efforts together to build a coffee shop corporation (Families for Psychosynthesis, Inc.), which sells stocks and provides therapeutic, recreational, occupational and financial opportunities for all family members. The family members have forgotten the "sick patient" role assigned to them by social convention and operate the coffee shop as a business in public. The family members *act* as president, treasurer, secretary, chefs, waitresses, managers, accountant or lawyer, and in a variety of roles in individual social relationships. As a by-product, the coffee shop over a one-year period has paid out $8,000 to the family members. This has enabled some families, otherwise unable to afford it, to continue in individual or family therapy. The income is derived from having the coffee shop available to the general public, thus linking the creativity and productivity of the families to the community at large. With the coffee shop, the private psychiatrist as a psychosynthesist liberates the latent power of his private practice as a community therapeutic force. The families' coffee shop has made the private psychiatrist a truly community psychiatrist.

In the Coffee Shop Mental Health Community, we are emphasizing not only the usefulness and unique contribution of each and every element of the eclectic psychiatry, whether it be psychotherapy, chemotherapy or electrotherapy; individual, marital, family or group—but above all, the union of all the elements into a meaningful therapeutic system.

Any competent eclectic psychiatrist can help to unite all his patients and their families to build such a system, which can double or even triple the therapeutic capacity of the solo psychiatrist. How can this be done? By telefusion. Telefusion is the union of several psychotherapy groups into a single super-group or mini-community by closed-circuit television. In this way, Michigan Institute of Psychosynthesis linked together more than fifty families in Lansing. In cooperation, the families built a coffee shop to satisfy the social drives and needs described by Adler in developing the individual personality in a social setting. The coffee shop provides the logical stage for continuing therapeutic processes. The coffee shop has become a dynamic theatre of psychosynthesis.

After telefusion couch, where does the eclectic psychiatrist start on videosynthesis toward the creation of a coffee shop mental health community?

Where does a psychiatrist begin?

Where did you begin with?

With the TV-linked Group Therapy.

The following three chapters will describe three TV-linked groups, which are carefully transcribed from videotapes to show how the three TV-linked groups were fused together to produce the TV-linked Families Coffee Shop as a center of Community of Mental Health.

9. FIRST TV-LINKED GROUP THERAPY

MORENO, FREUD, AND TV-LINKED FAMILY PSYCHIATRY

SYNTHESIST: Good afternoon, everybody. You were linked to the Tuesday Group, the APA meeting in Washington. You saw one of the foremost authorities on group psychotherapy such as we do here—Dr. Moreno. He is eighty-one years of age. Many things we are doing we learned from Dr. Moreno over the years. He uses psychodrama, in which you try to *act* a certain way in order to *become*. For instance, if you don't have confidence, you *act* like you are confident. You *act* like you are going to have a good marriage, then you have a good marriage. By acting this way, then you become. This is really common sense when you think about it. If you want to learn to speak French, then you have to act like you are a Frenchman or a French woman. If you keep on doing this for a few years, you become. And when you are not willing to act, then of course you could not get there. How else are you going to become if you don't act?

While I'm at it, I might as well tell you the story Dr. Moreno told me. You know, after eighty-one years of life, he's got a lot of stories to tell. When he first started on psychodrama, he said a woman came to him who wanted a divorce. She felt her husband did not show love for her, and she wanted a divorce, she wanted to die. So Dr. Moreno, who was young in those days, just said, "Well, if you want to die, why don't you die?" So she dropped dead right in front of his eyes. But Dr. Moreno, being a doctor, looked her over. She was not completely dead, because her heart was beating. She was in sort of a state of hysteria, and she got herself into this state of "death." So he called the woman's husband and said, "Tell your children your wife is dead." And, of course, the man almost burst into tears, and said, "She's dead? She just came to see you. There was nothing wrong with her. What did you do to her?"

So the man brought the children all dressed up and came to see Dr. Moreno, in this theater where he works—he is a psychodramatist—and she was laid in the center of the stage, like you lay the dead, all covered up with a white sheet. The children burst into tears and the man burst into tears, and they all wailed and cried because they saw the woman was really dead, as far as they were concerned, especially if the doctor said she was dead. After awhile in this state, she heard all the crying and then Dr. Moreno said, "Maria, wake up. You know they love you." She jumped out of the cover and was alive. It was a miracle, and the husband expressed all his love and his cares and his interest. She couldn't believe the children were so dressed up and her husband had gotten them ready all by himself, because she had always had to do everything like that. From then on, this woman stayed with Dr. Moreno in group therapy and became a child psychologist in Italy and is still going strong through that experience. She acted death, and the husband acted how he would really have acted if she were dead. He did care, but he just didn't know how to show it. Well, he acted he cared, and they got back together. Of course, that is only one episode. They have to keep on acting for a long time. So now you can see some of the things we are doing today come from a very wise man, Dr. Moreno. I notice you are laughing, Mary. Do you want to tell us something?

MARY ANN: I just thought that story you told was great. After all, that's what you went to Washington for, to hear other people, I guess. I haven't been doing too much, just sitting around the house. I'm going to work for my dad two days a week, on Friday and Saturday at his bakery.

GINA: I think that would be interesting to do psychodrama, where you all act a part. Let's see. I'm a little bit like Katharine Hepburn. Do I look like her? She speaks in a real husky voice, and she has freckles. Do I have freckles now?
MIKE: The synthesist uses psychodrama to a degree. That is what we do here.

PEOPLE DEVELOP IN RELATION TO OTHER PEOPLE

SYNTHESIST: Many things we learn happen right in this room. Moreno, of course, stands for group therapy, meaning the basic principle he is evolving is that people as personalities develop in relation to other people. When Gina talks about being a little bit like Katharine Hepburn, it is true. Katharine Hepburn is in her brain; whether she likes it or not it is in there. If she has a dream of talking to Katharine Hepburn, for example, that would be just as real to Gina as if she were talking to Katharine Hepburn in person, until she woke up. Or if she goes to a movie and watches Katharine Hepburn, Katharine Hepburn is not only out there on the screen, but in Gina's brain. She is very much in her. In synthesis, we know we are more of other people than we realize.

Freud is at the other extreme in approach. He is an individual analyst, looking at one brain to see that all the other personalities have altered that brain, namely, the mother, the father, and the relatives, people in the community. When Freud first came into practice, he was a hypnotist, he was a neurologist. In those days, hypnosis was regarded as a great discovery, but even that has been going on for hundreds of years. Very little we do anytime hasn't been around somewhere, sometime, in some shape or form, maybe not as developed. If you talk about a shot to the moon, as Edwin has told us, converting General Motors into building missiles and booster vehicles to go to the moon, you may think that is new. But the Greeks have thought about it long ago. If you know the story of Daedalus, he used wax wings and tried to fly, and the sun was so bright it melted the wax and he dropped dead. That's the moral to those who try to climb too high too soon before it is time.
EDWIN: I had an opportunity to listen to a doctor talking about alcoholism. He headed a committee that had gone all over the country, sponsored by the federal government, I guess. He went to different clinics and studied the people who had been to the clinics, and he discovered that physicians didn't want to admit that people were alcoholics when they came to the clinics. There were apparently at least four reasons for this attitude of the doctors not admitting people who were alcoholics: (1) If the person had insurance; (2) if he was married; (3) if he held a steady job; (4) if his drinking habits coincided with the doctor who was interviewing him. If these factors were there, the doctors for some reason wouldn't accept the patient as an alcoholic. This doctor said, oddly enough, it seemed as though the doctors would see just what

they wanted to see. He also stated that alcohol in this country will be here and the proper thing to do is accept it and teach people how to drink sensibly as they have done in Europe. I would like to have some comment on that.

THE LAW OF UNIVERSAL TRUTH

SYNTHESIST: What you are saying is, the brain tends to pick out things. If you look hard enough, you will find everybody is an alcoholic. But if you push the criterion higher and higher, then nobody is an alcoholic. It is a matter of use. Alcohol is used as a football, to fight the battles between individuals, whether between the law and the people, husband and wife, or parents and children. You use "alcoholic" as a label, so you can have something to beat each other with. Actually what is happening is the wills of people. People want their way, against somebody else's way. Freud said, it has got to be couch; he's got to be analyzed. Moreno says, it's got to be psychodrama; you've got to play everything out to understand what's going on. But in psychosynthesis, Freud is right and Moreno is right. If you recognize that thresholds, criteria, or definitions do come into play, then you understand the nature of conflicting views. If you change the definition, you change the picture. Now this is what we are learning: The Law of Universal Truth. You know the famous figure that was discovered by Jastrow (draws ambigraph on chalkboard). (See ambigraph, p. 160.)

Gina, what do you see there?

GINA: A duck.
SYNTHESIST: Okay, a duck. But if I say it is a rabbit, you would laugh at me.
GINA: Now I see the rabbit, too.
SYNTHESIST: You see the rabbit, too. How about you, Mary Ann?
MARY ANN: No, I can't see it.
GINA: Look at its bill.
MARY ANN: No, I don't see it.
SYNTHESIST: You just keep on looking. We will look at Mary Ann's face.
MARY ANN: (Studying ambigraph) Oh! I do see it. It's on the other side.
SYNTHESIST: Very good. Now you see it. I am just showing you how the threshold changes. The duck is a rabbit. But I don't see either one. You know what I see? Chalk dust. That's what we mean by the Law of Universal Truth. That anything you say is true. You could make any statement. There is a god, that's true. There is no god, that is also true. It is a word. It used to be called Zeus. Now if I said there is Zeus, you would think I am a crazy Greek. If you say there is no Zeus, you would think I am right. But 2000 years ago, if you asked the Athenians

whether Zeus existed or not—you'd better be careful! If you said otherwise, well, you know what they did to our friend Socrates. What did they do, Gina?
GINA: They didn't like his philosophy, so they said he would have to leave Greece. He said he didn't want to leave Greece, he would rather die. So he drank hemlock.
SYNTHESIST: But they also claimed he didn't believe in Zeus.
GINA: Oh, that's right. He actually believed in a god greater than Zeus, an unseen god. They saw Zeus as a human being; they thought their gods just lived above them, and had wars and fights and arguments, and they were like them, they were very human gods.
SYNTHESIST: Now you can understand people like Bertrand Russell, when he said he didn't believe in God. He was telling the truth, because his Zeus was different from God. Or what he believed was different from what other people believed. So we have come to the conclusion in this very room, over the years, that you can say anything as long as you do not do what? As long as you do not kill anybody. Because what good is killing, like the Catholics and Protestants in Ireland? What good is it that both sides claim they believe in God, but they kill each other?
GINA: That's not it. They keep that alive deliberately because of economic status. If both parties were very poor, they would unite as the poor in asking for better conditions.
SYNTHESIST: Yes, but the question is, if you ask them, they all say they are theists.
GINA: England wants it that way.

WHATEVER LEADS TO LIFE IS TRUE

SYNTHESIST: But we are not answering the question. If you ask each and every one of them, they will say that they are Catholics, they are Protestants, they believe in God, but yet they kill each other. Therefore, their conception is what Lord Russell said couldn't be true; what they believe couldn't be true. But Lord Russell is also wrong. You see, he did not learn the Law of Universal Truth. It is true, whatever they say, but where is the higher truth? The higher truth is, whatever leads to life is true. If it leads to death, it is false. Once you gain this universal truth, you can sit there and listen to anybody tell you anything and still keep your composure and a smile. If they say you are an alcoholic, you smile and say, yes, probably it's true.
ANNETTE: How could you be one if you didn't drink?
SYNTHESIST: It's not hard at all. You know what Dr. Moreno told us, you can act like a drunk, you can think like a drunk, and you can behave like an alcoholic. In fact, we just talked about this. Who is the worst alcoholic in the world? Think about him, and you become him. In your dream, you may dream yourself drinking, really enjoying the big drinks, under the table, pouring it in. You could identify. This is what psychodrama is talking about. You can act like one.

ANNETTE: But that's your imagination.
SYNTHESIST: Have you ever seen a great actor, who could take some tea in a drama and start to act like an alcoholic? If you watch him on television, you say he looks like an alcoholic; everybody has a laugh; you say, well, he's an alcoholic. It is the same thing Freud is talking about. The brain can carry these things out in one brain without physical activity, or you could carry this out with actual acting. To answer Gina's question, do you become able to speak French if you act like a Frenchman or French woman? The answer is Yes, but what she says is also true. She was changing the threshold. She asks if you can become French-speaking like that (snaps fingers). But the threshold was three to four years, if you keep acting like a Frenchman.
GINA: You would be a Frenchman?
SYNTHESIST: If you keep on acting. Act means you do things, you may go to France, eat the bread, drink the wine. Now the coffee shop downstairs is where Freud and Moreno meet. There is another man, Adler, a disciple of Freud, the first to rebel against Freud and believe in Moreno. Adler understands better than Freud did how a human being comes into social contact to become a human being. It's by having all the personalities in you. When you have older parents, say a man 50 and a woman 39, who have a baby, very often we find the baby is detached from other associates and not able to cope. The baby didn't have playmates and was always talking to the old parents. They are always doing things together, the three of them, so the child could not deal with life. There was fear of people. Okay, when you say you want to be very confident, you can work in the coffee shop and act like you are confident. Take Barbara, for instance. You wouldn't recognize her from her previous personality. When things are difficult, she acts confident, running a whole coffee shop in a medical center in a city. She couldn't imagine doing this a couple years ago, but now she carries on like nothing ever happened.
MIKE: In other words, when you start acting it, you build up more and more confidence and eventually you just take over that role.
SYNTHESIST: Right! That's where the coffee shop is psychodrama, except it's real. In Moreno's psychodrama, you actually come together and pretend or act it out. You remember one young man in the Friday group wanted to run a restaurant. With Moreno, you would ask the young man to come up here and pretend this is a restaurant, start to act. In psychosynthesis, we say why do that? Why not just go downstairs and have a part in the coffee shop and run it everyday? Then by the time you are 25 and have a little capital, you could start a coffee shop of your own.
GINA: I don't like to drive a car. I don't like machines of any kind, even washing machines. If I keep telling myself I am a good driver, and I don't mind driving in traffic, I will be a good driver?
MIKE: Why not?
SYNTHESIST: Right. In fact, they made studies of this, in Wiener on cybernetics. The brain is so very powerful. This man Wiener unites both Moreno and Freud in the cybernetic concept. They made studies of two

groups of young people. They had five people on one side who had never played basketball before, and they had five on the other side, matched according to age, height, weight, and other attributes, as close as they could. They told one group to practice basketball. They told the other group just to sit down and read about basketball and think about it. So the one group just thought about it, in the mind, even planning a sneak attack and other strategy, whereas the second group actually practiced basketball. Then the two groups were brought together. You know what happened? The group that thought about playing basketball got ahead, instead of the group that actually tried it out.
ANNETTE: Because they were concentrating?
SYNTHESIST: They were concentrating in their minds how to do it. This shows you could learn to do things by thinking about it, but then of course you have got to catch up by activities, they must act it out.
GINA: First of all, they did all the concentrating and then when they actually played, that was playing out the role, right?

SYNTHESIS COMBINES THINKING AND ACTING

SYNTHESIST: Yes. You see, one was on the couch, like Freud. The other one is group therapy. Then you put both together. What you thought, what you acted, and you put together. That's what the synthesis idea is. The idea is anything anybody says is true, even if it doesn't sound real to you or right, don't kill him. Don't get a knife and stick it in. What he says could be true, from his viewpoint. Just like I said, that is a duck, but it could be a rabbit. Don't be in a hurry. It took Mary Ann a few minutes before she could see it. But that's only one picture. A life is thousands of pictures. Now if you've got to convert every picture this way, it would take a long time. That's why therapy takes time. And that's why we need time in therapy. That's why therapy becomes very expensive if you only use professionals and individual therapy. That's where television and group therapy come in.
ANNETTE: I have a question. Would you say playacting is good, that everyone should do it, or should they just be themselves?
SYNTHESIST: It depends on your goal. If you are play-acting opposite your goal, you may be able to understand it, but if you are trying to play-act what you are trying to become, you keep on doing it, it becomes natural. You know, for some actors and actresses, their lives are more real on the stage than off the stage. In fact, they get very depressed when they are off the stage. They much prefer to be on the stage. Some of the television serials run for years and they become part of people's lives. It's more real actually than their life at home, fighting a little bit, a little pit of picnics, little bit of going to bed, you know, not very exciting. But the drama they see on television is pretty good, or in the movies, or on the stage, like Shakespeare's *Hamlet*, Tolstoy's *War and Peace*, Margaret Mitchell's *Gone With the Wind*, or reading *Journey of the Stone Monkey*!

Okay, have we put everything together? Let's put this in sharper focus. Moreno advocates psychodrama and group therapy, because he knows that the individual personality is a union of other personalities, and he also emphasizes the idea about father-figure. That means you need somebody who knows about the world to tell you about the world. You could not discover everything by yourself. If you tried to, it would take thousands of years. That's what the professional is for. I see families by the dozens every week. If you tried to fix up your own family, it would take you an awful long time, but we do this every day. It is like a garage mechanic. He fixes cars every day. He listens to the car, works on it, and knows what to do, if he is an honest one. Some are not too honest. Like one day, the meter that measures gas didn't work, what's it called?

LEE: Gas gauge.

SYNTHESIST: Well, it didn't work, so I took the car into the garage and the mechanic went through a big song-and-dance to show how complicated it is. But I was watching him. Finally, he said, "I fixed it." You know, all he did, he just plugged it in, in the back! (Group laughs.) And I said, you know, I saw what you did. And he was very honest and only charged $3.50. But another garageman, he might go through a whole thing and might say, I've got to charge you a professional fee, $25. So you can see where knowledge comes in. Knowledge helps you shorten the process. That's what a father figure is. But unfortunately today the father figure sometimes knows less than the son on some issues. He has not kept up with the times. And that's where you have conflicts. So it behooves people to keep on learning and keep ahead of their children in order to catch up with them. If you are just one chapter ahead, you will be all right. It is when you are one chapter behind that you get troubles. I just read my last chapter before I came in today! (Group laughs.)

COUCH SESSIONS VS. GROUP SESSIONS

SYNTHESIST: Freud, of course, concentrated on the individual because of shame and guilt. He capitalized on people's shame and guilt because they did not want to tell everything out in the open. Like Gina says, I don't want to tell about my problem in the group. So she needs couch sessions and they will be built for her. And Freud would gain in this respect. But then she has to pay for it, because the couch session is five times of a group session, 500 percent more she's got to pay. But if she's got a little bit more courage, she says to the group, my problem is this, let's talk about it here and find out what we can do. Basically, you can always get help either by couch or by group, but each has disadvantages. If you can't pay for something and, as a result, you can't have it, then what good is it as far as you are concerned? In psychosynthesis, then, we are trying to combine the best, most feasible, and helpful things together in order to make it work. That means you must take the best from everywhere and connect them up. That's how we developed TV-linked family psychiatry.

Do you mind if I mention one thing I learned from you and your husband, Gina?
GINA: No. Well, I don't know.
SYNTHESIST: Here comes the trust of the professional. She knows I wouldn't go beyond a certain point. She and her husband have different times to see me, because of timing, schedules, etc. I needed to know how to get them together. I can put her on television and I can put him on television and they see me together on television. This is what we mean by TV-linked family session, because without it, we would never get together. Am I right, Gina?
GINA: Yes. It was dangerous, too.
SYNTHESIST: She's honest! Jane?
JANE: I was reading about television in connection with a lady who had been complaining of chest pains while she was in an airplane. When they landed at the airport, they took her to a doctor's office and she never saw him in person, only by television. And he saw her by television and asked her all kinds of medical questions. He had a well-trained nurse who was with her, and he would ask the nurse to touch different parts of the woman's body, like he would have done, and he diagnosed her case as pneumonia. He was about two and one-half miles from her at the time, and he had her brought to the hospital and they checked her over and she did have pneumonia, as he had diagnosed. They figure if they can perfect this enough, it would cut down on the number of doctors that would be needed. If they can't get more doctors, it would be one way of getting more patients seen by the doctor.

TELEVISION AND TREATMENT

SYNTHESIST: Excellent. You see, what she is saying is exactly what we have been developing over the years. I've been listening to family members like Jane bring up new information to give us an idea. You know, television is not only useful for times when you couldn't find a doctor fast enough. It's also for times like when I was in Washington, you were still getting sessions. Some people don't understand, they think they've got to touch my hand and smell my cigar in order to get a treatment, that this is essential. But you need both. In psychosynthesis, they say, "I've got to see you, doctor," and that is true, but television can help also, and that is true. This is what we mean by synthesis. Some old-fashioned doctors used to treat pneumonia by holding hands, but the patient died. All the hand-holding didn't do much good. But when the doctor wasn't even there and the patient got a shot of penicillin, the patient lived!
ROY: Look at the monitoring system they've got with television now in intensive care in hospitals. The television is right there by the nurse's desk. All she has to do is look at the monitor and she can see everything you do.
JANE: With the lady who had pneumonia, they could even get the in-

formation from using the stethoscope and listen to it over television.
SYNTHESIST: Psychosynthesis is based on cybernetic theory. What the doctor wants to know is fed back to him by this wire. It is not really different than if I had gone into another room and spoken to you via TV-link from there and said, "I'm here in the room with you." I could then push a button, substitute with a relevant TV-tape treatment program, and go to see another family with an emergency. And that emergency family could turn out to be your own. The point is, with television, you have amplified the treatment power in helping the family, at a fraction of the cost. I will tell you a fact, 10 years ago, some of you couldn't afford the treatment you are getting today. A number of you in the room would not have thought about therapy because it was too expensive. Today you are getting more, and more intense, therapy at reduced cost.

Now, at the beginning of the session, we thought our title should be, Moreno, Freud and Psychosynthesis. But I think we should change that to Moreno, Freud and TV-Linked Family Psychiatry. That's even closer to our discussion, isn't it? Okay, see you next week.

VIDEOTAPES FOR PSYCHOSYNTHESIS

Groups for Psychosynthesis meet on Tuesday, Wednesday, and Friday afternoons. Each group is linked to the others through tevetapes. Topics discussed in May and June 1971 were as follows:

Tuesday Group

Date	*Title*
5-21-71	The Struggle of Two Personalities
6-4-71	Names, Identities and Family Relationships
6-11-71	Friends, Personality and Name

Wednesday Group

5-19-71	Medical Authority and Personality Conflicts
5-26-71	Separation of Families
6-2-71	Death of Mother and Fear of People
6-9-71	Television and Family Medicine

Friday Group

5-18-71	Transnomation: Impedance, Resistance and Synthesis
5-25-71	Trust, Truth and Therapy: Origin and Nature of ELT
6-1-71	Sex and the Nuclear Family
6-8-71	Conflicts and Contradictions in Identity

10. SECOND TV-LINKED GROUP THERAPY

TELEVISION AND FAMILY MEDICINE

SYNTHESIST: Good afternoon, everybody. I would like to introduce Dr. L. E. Harrington and Dr. Richard Hatton, family physicians in the Okemos and Lansing areas. Some of you may have known them. They are interested in TV-linked family practice, and we will try to answer some of their questions.

DR. HARRINGTON: I am acquainted with the TV demonstrations,

having taken some courses that your synthesist has moderated. But the use of television in group therapy or to present things other than basically psychiatric problems is relatively new, in my thinking. In fact, he first brought to my attention that certain aspects of medicine could be handled the same way, i.e., instructions to patients with various illnesses, like diabetes or weight control, new mothers and care of newborns, things such as that.

TV HAS IMMENSE APPEAL

DR. HATTON: I do general practice, and this type of television and communication with large groups has potential to help the family doctor cover a lot of things in a short period of time. And it has an immense appeal as a result. Basically, I would like feedback from you. We are specifically speaking about problems in general medicine, but you know the general family doctor has a lot of things come into his office that are emotionally oriented or psychiatrically oriented as well. We are interested in finding out from you how you would feel about something like this. Do you feel that this would be acceptable? Is the contact with the doctor enough so you feel you derive benefit out of this? Would you feel adequately handled were you to come to your family doctor's office with a problem and find that another group or three or four individuals with similar problems were there at the same time? And you were placed in a room and exposed to this type of program? Remember, of course, that after this is shown to you you would have an opportunity to discuss it either on a group basis or on an individual basis. The group would be even better for us, because we could cover more material with different questions coming up from each of you.
SYNTHESIST: Before we answer questions, shall we give them a hearty welcome? (Group applauds.) They are very busy practitioners and when they make a move, they do it with forethought and significance.
DR. HATTON: Let me ask Eileen. Have you been in the group very long?
EILEEN: No, this is my second attendance.
DR. HATTON: Do you feel this type of media has an appeal for you?
EILEEN: I'm not sure. I'm trying it out as an experiment. I was helped last time when I came because I felt maybe I could help Karen. She needed some encouragement and I did, too.
DR. HATTON: I would like to speak to someone who has been in the group longer and find out if they feel that the television and the things that have come out of it have been of value.
MIKE: We've been in the group about a year and a half and I consider it to be the best part of the whole deal. I get more out of this. With the doctor, I think a lot of times you go in with an ailment, and you don't get the needed time, whereas if you had a group situation, you could spend more time.
DR. HATTON: I want to get back to the television aspect, because this

is what appeals to me and Dr. Harrington. We are interested in knowing whether you as a group feel the tapes played from previous sessions have contributed to your betterment, have added to your knowledge of your problems, have enabled you to open up more and discuss more freely?

DR. HARRINGTON: Has it speeded up your treatment?

ANNETTE: I think the TV is of benefit to everyone because of the feedback on a person's old personality and things that have been disturbing them. Then when it is fed back to the group or to the individual, the person can profit because he can learn from the way he was before and how far he has advanced.

DR. HARRINGTON: You think feedback is a major advantage that television adds to group therapy?

ANNETTE: Yes, it gives you a better understanding of yourself and other people.

DR. HATTON: Why is that? Is it because of watching what has gone on previously you not only can hear but see, and you have a better recall then?

ANNETTE: Yes.

JOYCE: I agree. I haven't been in the group long and I've only seen a couple feedbacks of myself, but in that short time I have found out a great deal about myself, and I think TV is very beneficial.

DR. HARRINGTON: Were any of you in other groups without the television, before this one?

GLENN: I was. Of course, I had a lot worse problems than I do now. The TV gives a chance to communicate between the groups. We see the films of the different groups, and it gives you a chance to get their ideas on maybe a similar problem or the same one. There's a lot more information let out.

SYNTHESIST: How has the television helped your mother? Maybe they would be interested in that.

GLENN: My mother's from Battle Creek and only comes up here for the regular monthly meeting. TV tapes help her to see how I have progressed in that time.

TV HELPS US IDENTIFY WITH OTHERS

BARBARA: I think by watching group therapy on television we learn that people in the other groups have the same problems and difficulties, and from that we can identify with them, and it helps us in our situations and problems.

DR. HARRINGTON: Do any of you have suggestions for fields other than psychiatry, such as the things I mentioned? Does TV intrude on group therapy? Are you aware of the camera all the time, are you aware of the fact that you are being put on tape?

GLADYS: I think you could get a lot from a regular M.D. or general practitioner on television, like how to take care of yourself or other

people. But I think the TV has to grow on you.
DR. HARRINGTON: Does it take you quite awhile to get used to it?
GLADYS: That depends on the person. Some people grasp things quicker than others. I for one am a slow grasper. (Group laughs.) But the more I come, the more I want to come. I can't stay away!
RUTH: The TV has helped me several times since I have had ELT treatments, if I would get low, or my old personality would show. The doctor would show me a booster tape of what the old personality was and it would really snap me back to where I should be.
GLADYS: I think it might relieve this nursing problem, too, if some of these people from the homes could be shown how to do things. On television you could instruct people how to take care of the sick. They show mothers how to bathe the baby and all those things on TV now, as I understand it. For instance, showing it on TV wouldn't cost underprivileged people so much.
DR. HATTON: This is the type of comment I would like to hear, because I feel this particular media is definitely a part of the future of medicine. It's got to be. The number of people that a family doctor can see in a day is restricted, because it is a one-on-one situation. Where we can see several patients at a time, there has got to be a gain from that, and this is what we are interested in. But we want to know from you, having had experience with this, how you accept it and how you like it. And if you've felt that you've gained from it.
EDWIN: I would like to ask Dr. Harrington how TV would be presented.
DR. HARRINGTON: I can't answer you because I don't know. That's why I am here, to find out where and how I could use it, if it would be beneficial to my practice, and in what subjects it could be used.
EDWIN: As far as acceptance goes, wouldn't there be a psychological thing there? A family goes to their personal physician and wouldn't they demand his personal attention?
GLENN: It seems like you could teach a lot of skills on TV, show somebody how to do something with their hands, or something. It seems like schools would use it even more than they do.
MARY: I've been in a state hospital until about two months ago, and I think this group session is a fine thing. When you see yourself on television, it already gives you confidence. I think it is a big help, the best way.
JANE: I think I've benefited by television as much as any of the treatment. You can see how other people with the same problem or a similar one learn to cope with it, so that you, too, can work it out.
JOYCE: I agree that some patients would feel that this would be sort of impersonal because most patients want a personal contact with the doctor sitting in front of them and talking directly to them. But I think it also depends on the diagnosis of the patient. If the patient gets to know his diagnosis and his illness and has an understanding with the doctor that this is the way the treatment is going to be, then I think it would be all right.

DR. HARRINGTON: In other words, there has to be confidence in the doctor first.

JOYCE: Yes. Then I think the patient will accept the TV. If the patient is a diabetic, or has TB, or a heart ailment, this could be shown on TV explaining and perhaps showing someone else who has the same type of illness.

DR. HATTON: Do you think that a patient who has diabetes would feel better as a result of seeing a TV program where other diabetics are presenting problems, and then having a personal contact with the doctor later? Would he in turn feel that he would like to join a group? Do you feel this might be a possible avenue to follow as a result of the television?

FRANK: I think so, because the doctor doesn't usually have time to talk with you for a length of time, and if you want to know how the heart works, what's going to happen to you, what your treatment is, it would be nice to see it on film on TV. I know the doctor doesn't have time to do it.

RUTH: I think if I went to an M.D. and heard something like diabetes which was new to me and frightening somehow, seeing another tape on TV of someone else with the same problem would soften the blow a little more. You would see how other people live with this, and you would feel you are not the only one in the world. You could record this information to show and you could be doing something else with another patient.

MIKE: To start out, I think you are just going to have to take your patients and say, I have something here I want to try, and just gradually integrate it into your working schedule.

JOYCE: I think the doctor should have a personal interview first and then discuss the TV and explain to the patient that this is a form of treatment. I think the personal part should come first.

DR. HARRINGTON: The personal part should continue throughout, too.

ANNETTE: If you talk with the doctor about, say, diabetes, you hear it, but I think you remember more with TV, because of the sight. You could grasp it much quicker with the picture.

SYNTHESIST: Here is another interesting thing. Before Richard D. joined the group, he saw it on TV. What was your impression?

RICHARD D.: You really want me to tell you? I am going to be quite critical. I thought it was punk. It was very boring. Color would help a lot, but just about any TV I don't care much for. If you are going to put the patients on television, you could conceal the camera and the technician behind a partition or something, and it would make a lot of difference. To me, anyway.

DR. HATTON: If you knew they were there, even though they were concealed, would this make a difference?

RICHARD D.: I see what you mean, but it would make a lot of difference to me. I could relax more than I am right now. Now I am so high up I am not the least bit interested in it.

JANE: We just think of the cameraman as part of the group. Occasion-

ally we even get comments from him.
GLADYS: It has to grow on you, maybe three to six months, before you get rid of that tension. I was all keyed up tight at first, but now I'm just as relaxed as can be.
SYNTHESIST: Would you like to talk about Lona's problems, your old personality? I think family doctors like Dr. Harrington and Dr. Hatton would be interested in how we are helping people in the community.
GLADYS: I got all mixed up with the medicine I was taking. Now I am taking just the opposite medication that I originally took. The family doctor told me I was going to die and I believed him, but I found out that I am still here and everything is fine (laughing). In other words, I got all mixed up with the family doctor.
DR. HATTON: Do you feel television could improve or clarify relationships between a family doctor and a patient?
GINA: I would suppose so. It could improve understanding.
MIKE: I don't think it could be done right off. I think it would have to be a long-term patient who knew the doctor. You would have to have that personal relationship to begin with.
EILEEN: Doesn't it depend on the person? Some people have to feel closer to their physician than other people do, and they might not feel so close to him, but if he had a TV session and then had an interview following the TV session, I think it would be helpful.

TV AIDS UNDERSTANDING

DR. HATTON: Let's consider something like heart trouble. The doctor tells you that with the tests, the electrocardiogram, and the physical findings, there is a problem with your heart. Then he asks you to listen and watch a tape concerning other people with a similar problem. Would this enhance your understanding of your difficulty? Would it cement in your mind those things that the doctor wants you to carry with you for better knowledge of your problem?
EILEEN: Yes, I think so.
DR. HATTON: Do all of you feel this way? (Group nods.) In other words, this gives more communication, this gives more chance for you to have exposure to your disease state? This is something you would like to have?
JOYCE: If I was going to a regular doctor, I would like to have it.
EDWIN: I think some patients would think TV would be something like computerized medicine and the doctor would be treating everybody in a general way. You referred to heart patients, well, these people might be thinking, "My treatment shouldn't be exactly like this gentleman next to me here; he's got heart trouble, too, but I have a different kind of heart trouble, maybe." I believe it would be hard to make people understand.
DR. HATTON: TV in and of itself, then, is not the sole answer. You need personal contact as well.
EDWIN: Very definitely.

WAYNE: Ninety-nine percent of the families have nothing but good to say about the television. It is the feedback system that makes communication possible.
DR. HATTON: I would like to hear anything about television that you do *not* like. Anybody? (Group is silent.) So regardless of what the problem is, you would have no major compunction to taking your time to be part of a group if you felt that you could help somebody as a result?
JOYCE: The only thing I don't like about the TV is that sometimes I don't concentrate and take in as much as I would like to take in. Later I think, if I could just see the tape over again or if I could discuss it with the doctor or someone else, it would be more beneficial.
DR. HATTON: But the fact that this is on tape means it could be repeated, so if you did want to see it again, you could come at another time.
MIKE: What would you do about clarifying a tape? If I had questions to ask about a tape I've seen, how would you handle this?

TV AND THE DOCTOR-PATIENT RELATIONSHIP

DR. HATTON: This is an area where we would have to make a definite commitment in advance, so the patient would know that although we were using this new type of media, in an effort to expand on a problem he has, that the doctor is still definitely at his disposal, definitely wants to talk to him as a patient, and definitely wants to continue to know him as a patient on a one-to-one basis. The TV tape and the material presented on it are simply an effort on our part to make him more knowledgeable about his disease and have a better understanding of what we expect out of him in an effort to help with his disease.
SYNTHESIST: We are going to present some of this material to the American Medical Association, as to what we have learned in this room, and the television techniques. Your comments are very important to physicians all over the country, not only in Lansing itself, in order to make communication improve between doctors and families and patients. We are interested not only in disease but in health. We should emphasize how we can prevent disease from coming on. Now you two doctors have seen the system, which we have had in full operation for three years. I would like to ask you a few questions. First, what is your general impression, having seen the families and the TV system in operation?
DR. HARRINGTON: Well, it obviously works. You certainly wouldn't have used something that was a failure for three years. I hope not! The problem I have to answer is if it would be beneficial for my practice in areas other than this, how beneficial it would be, if the TV could help the person or do something I couldn't do or wouldn't have time to do. I think there are some things—teaching programs, for instance, where it would be very helpful.
SYNTHESIST: Dr. Bonta Hiscoe is joining us now. He is a surgeon and

president of the Ingham County Medical Society. We are glad to have you with us. The second question, Dr. Hatton, is: If you were setting up such a system, what would be the first way you would want to use it?

DR. HATTON: I would take problems in general medicine as a kick-off point, because my familiarity with them and the type of practice I have would lend itself to building up the system more rapidly, branching out into areas of emotional difficulties and family problems. I think it would be beautiful if I could establish a working relationship with someone such as you, a psychiatrist, who has a similar type of system, where tapes could be exchanged and help could be gained. I could easily visualize a doctor in a rather remote area, where he doesn't have access to specialists, putting this thing to a fantastic use, if the other physician to whom he is communicating has a similar type of system.

DR. HARRINGTON: Tape consultation.

DR. HATTON: Tape consultations with surgeons, with internists, with Ob-Gyn, with psychiatrists, there is just no end—it's a complete new vista.

SYNTHESIST: You could use TV not only for self-visualization, but you could tape the whole examination and send it to the specialist.

DR. HATTON: Absolutely.

SYNTHESIST: How would you change your practice, then, if television were introduced?

DR. HARRINGTON: I would have to change the physical layout first. I'd have to have a larger room than the examining room I have. Then I would have to establish groups for various things. For instance, patients with heart disease, or diabetes, or cancer; patients concerned with problems of aging, or rearing children, or many subjects like that, could be organized into groups.

SYNTHESIST: Here's a question I think many patients and doctors would be interested in. Do you think you would be able to lower the cost of the visit to the office?

DR. HATTON: That's a very important question, and I feel in many instances, yes. General problems of medicine require a fantastic amount of time, for which the patient has to pay. For example, I sit down and take an extremely long time talking to diabetics about their care of their body, their diet, the way they give insulin, etc., or talking to heart patients about various aspects of their condition. But if I could put that on tape and have half a dozen or so people come at one convenient time and sit down all together and see this thing, why, the cost would be considerably reduced. You can anticipate that the patients would have questions after the tape, but still your time is cut down, because then you are answering specific questions, not generalities. This is what makes this thing so appealing.

DR. HARRINGTON: It's obvious it would cut the cost. The only thing the doctor has to sell is his time, his brain, his knowledge. He certainly doesn't have a product, and he can only utilize that in the number of hours in a day or the number of hours he wants to work. If he can get an idea across to seven people in the same time he would have been getting across to one person, it would be an extremely rare doctor who

wouldn't pass that saving along to his patients.
SYNTHESIST: Do you think amplifying the TV-linked system would solve some of the manpower problems?
DR. HATTON: I think it has the potential to do so, but remember that to make a tape requires time and effort and study, that the tape has to be updated, that you can't sit on a tape that has been made five years ago and keep showing it, you have to keep the material current. But I do believe that if you counterbalance that time and effort with the amount of time you would have to spend with just one patient, the overall time is a saving for the doctor. The initial cost outlay for the equipment and so forth is something that has to be considered, but I think that like any piece of lab equipment, it is going to pay for itself over a period of time.
SYNTHESIST: What help would you need in setting up such a system in family practice or a specialty?
DR. HARRINGTON: You would need a good cameraman and somebody who knew something about operating the machines. Is that what you mean?
SYNTHESIST: Yes, how would you go about it? If you decided you wanted to start, what help do you need?
DR. HISCOE: Money. (Group laughs.)
DR. HARRINGTON: I'd probably come back here and find out how to set up the group!
DR. HATTON: Well, I would utilize the information I could garner from somebody who already had a setup functioning. Your ingenuity will determine how much you are going to use it and what you are going to do with it, but to get the beast in shape, so to speak, I would consult with people who are already using it and also take advantage of the knowledge of the people who are in the field.

TV AND THE MEDICAL MANPOWER PROBLEM

SYNTHESIST: Perhaps Dr. Hiscoe can comment on this. What is the medical manpower problem?
DR. HISCOE: I think everybody knows there is a tremendous amount of pressure both on the local and the federal level for some kind of change so we can cover more people more efficiently. And every technical help as well as every new personnel that we can use will be an improvement. The use of electronics is a very popular thing. I think particularly computers and so forth at the moment have become the vogue. I think the use of television is obviously a very worthwhile and useful technique as well. I think ideas from practicing physicians will give us some of the good answers. They do know the problems of the front-line, and hopefully they will give us some things that are practical rather than theoretical.
SYNTHESIST: What kind of front-line problems are you talking about, in terms of manpower?
DR. HISCOE: Obviously, as you say, we are short of doctors, which is

one problem. There are those who say there is not such a shortage of doctors as an inefficient use of doctors and the personnel that we have today. And I think television is a system to increase the scope of a particular physician with his patients. He can cover more. I think in conjunction with this, if you were to involve some type of paramedical personnel to run the show—this would be one of the cost factors that you were talking about, too—hopefully, the doctor wouldn't really be responsible for the technical details of getting it to the patients. His nurse or aide could do this and save him all of these details. This is obviously a cost, but I think it would be less than if he did it himself. Another way it would be useful is for preoperative patients in the hospital. The nurses on each floor have conferences with patients who are going to surgery. A tape such as you are speaking about would be useful to give them general background, at which time the nurse could answer questions specifically.

SYNTHESIST: Do you think some of the problem in doctor-patient relationship has been due to the lack of time for communication?

DR. HISCOE: Right, right. I think television is a reasonable substitute, better than lots I can think of. I think you are still going to need some personal time, however. I don't think you can eliminate it completely.

SYNTHESIST: Here in this clinic we use both. In other words, first I present myself to relate to the family and the patients, and then I choose the taped subject matter that is relevant to discussion. After the tape has been shown and is finished, I come back to write prescriptions, or give medication, or give other instructions, and get some feedback. Are you convinced this has real possibilities for amplifying manpower and improving the doctor-patient relationship and practice?

DR. HATTON: I think the potential here is just very great. I think the relationship between the doctor and the patient can always be improved when there is more contact. Even if it isn't a personal contact, if the patient has a chance to see and hear something that the doctor has produced, it is kind of like a prescription. You take a pill, you take a TV tape. The connotation is to some extent there, I think, that the doctor is doing something for you that will help you, and for this reason I think there is a very definite potential.

SYNTHESIST: We do this very thing in the clinic. We write a prescription that says "TV booster session on child-rearing" for a certain family. They will come and follow through a TV taped series on child-rearing. For a surgeon, it may be preoperative techniques or post-operative exercises, whatever you prescribe. If this is the direction we should move, how can we convince the physicians? How can we make this knowledge available as we have experienced it here in Lansing, Michigan, Dr. Hiscoe?

DR. HISCOE: Well, the method of communication in medicine is fairly well standardized. I mean it is quite ubiquitous. There are ways of communicating through journals and certainly there are enough throw-away publications in this day and age. There are conventions such as you are preparing for, and I guess word of mouth. Satisfied customers are still the best advertisement there is.

SYNTHESIST: Does anyone have a question for the doctors today before we conclude the session?
JOYCE: I would like to know which one of you will be starting this first! (Group laughs.)

VIDEOTAPES FOR PSYCHOSYNTHESIS

Groups for Psychosynthesis meet on Tuesday, Wednesday, and Friday afternoons. Each group is linked to the others through tevetapes. Topics discussed in June and July 1971 were as follows:

Tuesday Group

Date	*Title*
6-8-71	Conflicts and Contradictions in Identity
6-15-71	The Responsibilities of a Synthesist
6-22-71	Synthesizing New Group Members
6-29-71	Families, AMA and TV Technology
7-6-71	The Anniversary Phenomenon
7-13-71	The Futuristic Family and Marriage

Wednesday Group

6-16-71	The Creative Solution
6-23-71	To Spank or Not To Spank
6-30-71	Fear, Memory and ELT
7-7-71	The Marital Warfare
7-14-71	The Art of Limit Setting in Life

Friday Group

6-11-71	Friends, Personality and Name
6-18-71	The Creative Formula
6-25-71	Feedback, Pledge and Synthesis
7-2-71	Help, Self-Help, and Help Others
7-9-71	Limit Setting in Family Life
7-16-71	Concentric Expansion of Communication

11. THIRD TV-LINKED GROUP THERAPY

FIRST TV GROUP BOOSTER

WAYNE: You are probably all wondering where our synthesist is today. An emergency came up, and he couldn't be here. I think as the tape you watched indicated, the idea of continuity is important. Because the synthesist had to be somewhere else today, I don't think the whole group should leave. The synthesist felt that for continuity the group should go on and meet just as though he were here. So let's get started. Who wants to talk first?

GLEN A.: It took me almost half the program we watched to dig out the definition of continuity. What I think it means is a continuation of the service that is rendered. If you've got a goal, you've got to have continuity to keep the service going to meet your goal.

CONTINUITY AND GOALS

WAYNE: If I read you right, you are talking about continuity in terms of a goal. It's to continue to pursue your goal and not to deviate.
LORRAINE: I agree with him.
WAYNE: What's your feeling on the services continuing even though the synthesist isn't here?
LORRAINE: They should continue.
JAY: I don't believe it is just services, it's all life that should continue on, not to stop it just because of one interruption. It should be kept going. Goals are okay, but you have to keep going even if they get interrupted.
WAYNE: Then you are saying you can deviate from your goal, but as long as you know where you are going, you will continue. Services don't depend on one person, they are continuous.
BARBARA: I have two ways of continuity, as a wife and mother and also as a part-time worker in the Coffee Shop. I need both ways. By the way, I talked to someone today who asked me if I would mention the fact that at the forum Saturday the principal of Sexton High School will be the speaker. Everyone is urged to attend.
WAYNE: So continuity occurs in many areas. What is your reaction to the group continuing?
BARBARA: In many cases, and this is just one, because one person isn't here doesn't mean that things can't continue on. In fact, they should. Because the doctor isn't here doesn't mean that everything stops. They must go on. We all need group therapy.
WAYNE: You feel you are not dependent on the doctor, you are interdependent.
LINDA: I've got a problem. My mother lives in Ionia and she *never* calls me, but last night she did. Since I've had ELT's I don't depend on her as much as I did, and I think she feels awfully hurt and lost. I don't know what to do because I don't want to hurt her. I want her to think I still depend on her, but I don't want to take things from her.
WAYNE: May I ask what you mean by "I" used to depend on her?
LINDA: Teresa used to depend on her.
WAYNE: What would be your reaction if you heard the synthesist was going to be out of town and couldn't be in the group but the group was going to continue?
LINDA: That's all right. If he has to go, he has to go, and there is nothing you can do about it. As long as everybody is here, you might as well go on.
WAYNE: You think the group not only depends on the synthesist but on the other families also?

LINDA: Right.
LEE: We can't depend on the synthesist as a crutch. I think we all can carry on quite well without him. Once in a great while it doesn't hurt to carry on without him.
WAYNE: The families have risen to a level they can depend on themselves, interdependence, you are saying. What was your reaction to your wife's problem?
LEE: I'm glad Teresa's mother spent the money on a long-distance call.
LINDA: You're not very nice. (Group laughs.)
LEE: That's the truth, that's the way I feel. I think it's about time she started using the telephone and calling us for a change.
LINDA: You've got to admit that's the first time she ever did.
LEE: Well, I'm proud of her, I really am. (Group laughs.)
MARY M.: My mother and father and sister come here and I would like to stop seeing psychiatrists. I mean I say it over and over again and it doesn't do any good. I'd like to stop seeing a psychiatrist. That's all I've got to say.
WAYNE: How do you interpret this in terms of continuity?
MARY M.: It's just that I say it over and over again, to please stop talking about seeing a psychiatrist and they say no, you just come on and you see one anyway.
WAYNE: You are saying the need for a psychiatrist should advance to a psychosynthesist, who you are interdependent on instead of dependent on.
MARY M.: Yes, I'm dependent on it but I would like to be independent too.
WAYNE: Interdependent.
MARY M.: Interdependent.
WAYNE: This is your first time in Wednesday Group. Let's give her a welcome. (Group applauds.) And there is another new member. What is your name and what do you hope to gain from the group?
GINA: Gina. The synthesist asked me to come here.
WAYNE: What's your reaction to the tape we saw?
GINA: About continuity? I agree with that. I mean some things. I don't think everything should be continued.
WAYNE: Do you think the group should go on even if the synthesist isn't here?
GINA: Oh, yes. They seem capable.
WAYNE: Let's give Gina a hand too. (Group applauds.)
ROY: I think Glen had continuity pretty well defined, but you could go one step further. You can't accomplish anything alone. You've got to have people in order to accomplish anything. That has been proven here right in these group sessions. We alone have our goals set, but we need help from the rest of the group to reach this goal, and that's the reason for the continuation of the group even though the doctor isn't here.
WAYNE: In other words, no man is an island, men are social beings and need other people around them. This interdependence between the

families and the synthesist is made possible by the families working together toward a common goal.
KEN: I've been trying to take my second set of ELT's. That was supposed to be my fourth one and I want to get out of them. Just like that. For one thing, they scare the s--- out of me.
WAYNE: Do you want to define what you mean by "I"? Which personality?
KEN: Ken, I guess.
MIKE: You've got to want to take them and know you need them yourself. No one else can tell you.
ANNETTE: You've got to have confidence in them.
MARK: I know I need them, but I just can't keep up with it.

OLD PERSONALITY RESISTS ELT'S

WAYNE: Mark, do you see any relationship between resistance and the old personality or the new personality? Is there any tie-up there at all? I mean, how are the ELT's in relation to your old personality?
MARK: I don't know. It is still my old personality. I've never been a new personality yet.
GLEN A.: Do you feel more calm?
MARK: No.
MIKE: Does anybody else see any difference in him?
MARY H.: I do. I think as far as continuity is concerned, it seems kind of funny without the synthesist here. I think we need him. It seems real different without him, but I think everybody can get along. Life must go on. And as far as Ken and his ELT's is concerned, Mark's old personality had a lot of trouble with headaches and back aches and chest pains and every other kind of pain when he was going through them. He thought he was going to die off and all that; they were bad, you know. But I have something else to ask here. We're in the process of buying a house. Ours is sold. At the time, we figured we had enough money for the house and now all of a sudden we need a thousand extra dollars. The budget is strained now so no more could be borrowed for payments. I just wondered if they sue you, or what? We haven't signed the final papers, but we signed the papers to buy. Unless some big miracle happens, we don't have the money. I was wondering if anybody had any good suggestions.
WAYNE: Do you have any ideas at this time about continuity and the TV? How the TV could be related to continuity?
MARY H.: It's a continuing process. I don't quite understand what you mean.
MIKE: The TV allows you to go back and allows you to go forward. If the synthesist is not here, you could put on a tape with him. So he can go to Timbuktu if he wants and we could still have him here. In other words, he could tape his comments for today. He couldn't answer

questions but he could tape what he wanted to comment on.

MARK: The biggest problem is that he is not here at the end of the group session to recap. Without that, all you've got is what you have gained from everybody talking. If you don't bring out enough, you have wasted an hour.

WAYNE: I see two things there. The idea that through TV technology a person can be here and not be here, and the possibility of wasting time. Glenn, how do you react to continuity?

GLENN B.: Continuity to me seems like keeping things the same, at the same pace, not making any big drastic changes. Having no group at all today would be kind of drastic. It would be a big change from the normal. I think that keeping things the same is what continuity means.

WAYNE: How do you feel about the group in relation to the synthesist not being here yet?

GLENN B.: I suppose once in awhile it would be okay, but like Mark said, without the comments toward the end it is not as helpful. I think we need his remarks on things, one way or another.

JANE: I feel continuity is a continuation of something, and I suppose we can get along without him being here, but I like to hear his comments at the end of the sessions also. He speaks so easily and so understandably that you really get quite a lot out of what he has to say.

WAYNE: Very much like yourself. Gladys?

GLADYS: I haven't been here for a couple of sessions, so I kind of lost out, but I think it is better to continue. I found that out after being away for a couple of weeks. I didn't do it intentionally that way, but I missed it. If you stay away, there's a break there and when you come back you feel a little bit on the lost side.

WAYNE: Anything else you'd like to bring up?

GLADYS: I'd like to know how everybody controls their emotions.

ALICE: I think we are all still learning that, and the synthesist is a good teacher. Our continuity rests with the group. We have continuity of thought from the TV tape we saw at the beginning and the group discussion, and we hope the discussion after the group is finished is part of the session.

WAYNE: Then we can gain continuity from each other, also?

MIKE: If the synthesist isn't here at all, you will lose a certain amount of continuity because everyone brings out certain problems, comments, and so on, and you need the synthesist's viewpoint. Like Mark and Mary and their problem with buying a house and what could happen. The synthesist usually comes up with a good answer on something like that. You miss it when he is not here. You've got to have him here at some time. I don't think he intends to get away from that, because continuity is like a chain. You take away one link when you are pulling a car, and that is it. You are not going to pull it anymore.

TELEVISION AIDS CONTINUITY

WAYNE: Our synthesist was concerned that he couldn't be here today, but as many of you know who have family sessions, he can be in many places at the same time, through the television. That's why I purposely asked that question. As far as Mark's question, and answering specific questions, the synthesist will see this tape and can answer questions that have been raised.

KEN: Back to my problem. My wife and I talked it over with the synthesist, and he was talking about couch sessions again, to probe deeper. I'm trying to answer my own question of should I go on with ELT's or should I wait until he gets all the answers.

ROY: Did you go back to your old personality again?

KEN: I blew my cool a little bit. During the strike when I was thinking about money and all that, I did revert back. I saw that I needed some more ELT's but he told me I had to take ten, and I didn't want to, even when I thought I should take some.

ROY: You saw the change that was made in the first ten, but if you reverted back again, the next ten will probably bring you over the top to reach your goal. You asked if there were changes in Mark. I think so, because he used to come in here smiling and sociable, and today he's reverted back to the old personality. I have been coming here a long time, and I have seen changes in people that are almost miraculous. There's a couple right there that is proof of the pudding (indicates Mark and Lorraine). I saw that all the way through. They were about ready to separate and break up their home and now I think they are very happy and doing a good job raising their child, and they seem to have a good communication between them.

LORRAINE: People who have ELT's expect things to happen just like that (snaps fingers), but it takes time. You have to have confidence in yourself and you have to really push yourself to obtain the goal you are trying to reach. But if you think, "It's going to change me completely and now I am going to be this person," well, it isn't going to happen that way. You have to push yourself. With Mark and Mary, it could be because they need the extra money for their home, and that could be depressing Mark.

MARK: I agree. People expect too much to happen right after the ELT's. *Exactly* right after. It doesn't happen that quick. I don't know if the synthesist has ever said how long a time. He probably can't put a time on it exactly, because each individual would be different, but I think he would agree it is a month or two after the last ELT.

LORRAINE: Because your mind is loosened and it takes time to obtain things. Your old personality is still there, but you have to be strong enough to fight it.

ROY: Those of you who have had ELT's, I have seen a change gradually, a little at a time, a little at a time, where you could see the change coming.

MIKE: The change comes from talking to the synthesist in sessions,

family sessions and from here in the group. He has mentioned that it usually take some years for a regular psychiatrist to get any results. If you retain anything at all from talking to him, you will find yourself adjusting your life. When you talk to people and meet people and do different things, you will find that what he has said sticks in the back of your mind. You get so that different things you have learned from him actually guide you through your life, from day to day.
JOHN: Because the synthesist is not here yet, we are going to show another tape, to keep continuity.
(Group watches tape of another session. At its conclusion, the synthesist enters the group room.)
SYNTHESIST: Hello! You have participated in the first TV group booster session. I was not able to be here because I had to be in court. Today's experience is important. We didn't plan it this way, but it happened. I want to get a response from the group. Would you like to maintain stability and continue next week the same way so that you could learn to be interdependent with the families, when the professional may have to go elsewhere? This will happen next week when I am going to Washington, D.C. to present tapes of the families and what we have done here to help other families elsewhere. But I will be here with you on television and giving you information. Could we see a response, show of hands, as your reaction to TV booster session? (Group members raise hands.) Okay, this will help other families and psychiatrists in Washington.

TV GROUP BOOSTER CUTS FEE

SYNTHESIST: For the TV group booster sessions, you will hear that for the first time in your life, the doctor will lower his fee. For all these sessions, it will be half, $5 instead of $10 per person. You may have noticed recently you are paying to MIP, Michigan Institute of Psychosynthesis. What we are trying to do is maintain stability for you, your family and your children, for years to come. When I decide to go to Florida to lie in the sunshine, your children may be here and we keep their records so you could maintain family stability. Sometimes families go on vacation but I still have to be here. But when I go on vacation, I do not see any reason why people who need help should not continue to have help. So you were really making a historical record, the first TV group booster session, while we were in court today. Do you have any questions before we conclude? We will meet here next week for a TV group booster session, and we will reduce the fee, as I mentioned.
WAYNE: What we are going to show next week is your booster session, to show feedback on ourselves and how we have grown over time, and also a sneak preview of what we are going to show the psychiatrists in Washington. But can you imagine what we have succeeded in doing? To bring psychiatric fee to five dollars a session and to have continuity when it is not usually available. From there we can expand to several directions, as we wish to.

THE CASE FOR PSYCHOSYNTHESIS

SYNTHESIST: This is the *case* for psychosynthesis: C for continuity; A for accessibility; S for stability; E for economy. I shall see all of you in Washington, because we are going together, on or off TV.

VIDEOTAPES FOR PSYCHOSYNTHESIS

Groups for Psychosynthesis meet Tuesday, Wednesday and Friday afternoons. Each group is linked to the others through tevetapes. Topics discussed in April and May 1971 were as follows:

Date	*Title*
	Tuesday Group
4-20-71	Synthesis, Yvonnie in Africa and in America
4-27-71	The Case for Psychosynthesis
5-4-71	First Tuesday Group Booster
5-11-71	Elements, System and Programmer
	Wednesday Group
4-21-71	State, Families and Coffee Shop
4-28-71	First TV Group Booster*
5-12-71	Moreno, Freud and TV-Linked Family Psychiatry
	Friday Group
4-16-71	Marital Stability and Brain Function
4-23-71	Continuity and People's Sentiment of Service
4-30-71	Merchants of Information
5-7-71	Continuity, Commitment and Feedback
5-14-71	Psychodrama of Divorce

12. TV-LINKED CYBERNETIC SYSTEM OF PSYCHOTHERAPY

> *There is therefore nothing surprising in considering the functional mental disorders as fundamentally diseases of memory, of the circulating information kept by the brain in the active state, and of the long-time permeability of synapses.*
>
> (*Cybernetics, 1961*)—NORBERT WIENER

Norbert Wiener[8] had extraordinary insight into the basic problems of modern psychiatrists. The above quotations reflect clearly his prophetic vision on the probable course of modern psychiatric development. He was moving very closely toward psychosynthesis when he further speculated that: "It perhaps explains, too, why there are circumstances where a joint use of shock treatment and psychotherapy is indicated, combining a physical or pharmacological therapy for phenomena of reverberation in the nervous system, and a psychological therapy for the long-time memories which, without interference, might re-establish from within the vicious circle broken up by the shock treatment." This statement may be regarded as the earliest and clearest formulation of cybernetic psychiatry, fusing electrochemotherapy with psychotherapy as a psychosocial information feedback process. This is psychosynthesis[5]. And the psychosynthesis model is cybernetic in conception.

HISTORICAL REFERENCE

Freud (1886) was disappointed in Erb's electrotherapy and abandoned it for hypnosis before developing psychoanalysis[1]. As electrical engineering advanced, Cerletti (1938) restored electrotherapy in the form of electro-convulsive therapy. Now (1969), we have come about a complete cycle to where Freud had originally departed from Erb's electrotherapy. Psychosynthesis is, therefore, the child of psychoanalysis and cybernetics[6]. And we are intensifying psychotherapy with electrotherapy.

Since 1960, as a practicing psychiatrist, I felt the need to develop a unified theory of psychotherapy as speculated by Wiener in order to answer some of the basic problems and needs of patients and families. As you probably heard and read about the rising criticism against psychoanalysis and the competing claims of all schools of psychotherapy, chemotherapy, electrotherapy, behavior therapy, etc., the debate has been quite confusing and bewildering for all concerned. In 1961 the author[6] began to follow the cybernetic concept to unify all opposing theories by uniting the elements of psychoanalysis, psychotherapy, chemotherapy, electrotherapy, behavior therapy, family therapy, group therapy, and community mental health programs into a single television-linked feedback system. In the current development, psychosynthesis has become a TV-Cybernetic system of psychotherapy. However, to construct and to unify these eclectic elements into the psychosynthesis model, we require three basic conditions to realize this system in actual practice.

THE THREE BASIC REQUIREMENTS OF PSYCHOSYNTHESIS

To construct the psychosynthesis model it is necessary to achieve selective or differential memory erasure and subsequent reprogramming. If we can achieve selective or differential memory erasure as hinted to by Wiener[8], then we are on our way to build a scientific system of psychiatry. I believe we have the basic requirements even though not in the most streamlined forms. Future research may be goal-directed to improve these possibilities, which are definitely available for the construction of a psychosynthesis system. The three basic requirements of psychosynthesis are as follows:

1. The phenomenon of retrograde amnesia;
2. The technique of memory reactivation;
3. The procedure of immediate reprogramming.

1. *The Phenomenon of Retrograde Amnesia.* This phenomenon has been known to neuropathologists and neurophysiologists for some time. It is a loss of memory extending back over a period from the onset of disruptive mechanical or electrical force to the central nervous system.

In short, memories of most recent origin will be erased first by sufficient physical or electrical force, to cause the loss of consciousness. I define *consciousness* as the most recent memory or scintillating memory. We have, also, a circulating memory and a permanent memory as described by Wiener[8]. The phenomenon of retrograde amnesia provides us with a neurophysiological basis for selective or differential loosening or erasing of scintillating memory, circulating and permanent memories.

2. *The Technique of Memory Reactivation.* Memory (consciousness) reactivation technique is related to hypnosis, suggestion, and free association techniques. This may be regarded as a modified form of psychoanalytic approach of Freud.[2] Permanent memories or childhood

memories can best be reactivated by free association on the couch. However, the author has improved on the technique by recording couch productions on videotapes, which are subsequently used for memory reactivation. This technique provides the needed condition for differential memory loosening and erasure when used in combination with phenomenon of retrograde amnesia.

3. *The Procedure of Immediate Reprogramming*. This immediate reprogramming is carried out when the patient is actually in the state of unconsciousness, and when kinetic and circulating memories are in the loosened state. The procedure of reprogramming can also be regarded as a phenomenon of neuro-physiological "over-riding" as pointed out by Warren McCulloch.

Now with the above three conditions satisfied, I shall classify the types of memory in order to give some insight as to the nature of psychosynthesis and the direction of our future work. Memory can be classified as follows:

(A) Static Memory

1. Permanent memory, genetic information of evolution, probably encoded in DNA and other protein molecules, which are relatively unaffected by the electrolytic therapy.

2. Semi-permanent memory, experiential information from childhood onward (e.g., Oedipus complex, language, etc.), probably encoded in RNA and other protein molecules, which may be altered somewhat by the electrolytic process. However, future research may yield more effective electrochemotherapeutic agents. At the present, phenothiazines are used with the electrolytic therapy to induce lysis of memory.

(B) Kinetic Memory

3. Circulating memory, continuous background input-output information between neurons and neuron groups with lowered threshold of recent events, places and persons, language and social relationships, which can be markedly loosened or erased with retrograde amnesia for several months by the electrolytic process (e.g., the patient forgot he started a divorce proceeding against his wife two months prior to ELT).

4. Scintillating memory, or consciousness, is the moment-to-moment integration of information of permanent, semi-permanent and circulating memory together with current input information of reality, (i.e., via the eye, ear, etc.). The scintillating memory or consciousness is dramatically loosened and erased by the electrolytic process. This is self-evident since the person becomes unconscious undergoing the electrolytic process of psychosynthesis. Consciousness is self-creating, and ELT produces a psychological mutation, a psycho-mutation of information, so to speak. This permits the reprogramming with fresh informa-

tion for the synthesis of a new personality post-lytically, as explained by the psychosynthesis model.

THE PSYCHOSYNTHESIS MODEL

The psychosynthesis model can be represented as follows:

$$E_i(E_s) \rightarrow E_o.$$

E_i = input energy pattern. E_s = stored energy pattern of genetics and experience. E_o = output energy pattern, i.e., behavior. A personality is defined as a "Holographoid matrix," scintillating continuously to E_i by evoking probabilistically the elements of E_s as if from a videotaped memory bank to discharge E_o which is fed back into the system to recreate consciousness. And consciousness is neuronal scintillation. The Gabor[3] Model of hologram may provide a theoretical basis of selective erasure. And the energy function of E_i, E_s, or E_o is the same E as in the mass-energy equivalence equation of Albert Einstein: $E = MC^2$. The author [8] views mind as mass in relative motion and brain as energy at relative rest. Consciousness is a single frame of electrical charges in motion, like electrons bombarding a television screen. And personality is a time-series of this scintillating frames of consciousness. Personality, therefore, becomes a reverberating input-output pattern of self-creation, seeking information or patterns of energy from the environment as well as from its own memories. The stability of any given personality is its identity, which is maintained by feedback upon *the principle of most-similarity*. In other words, the personality never recreates itself but only a pattern most-similar to itself. The phenomenon of unique individuality and personal continuity depends on memory. The psychosynthesis model thus provides us with a clue as to how a personality may be radically transformed. The Cartesian dictum of "I think, therefore, I am!" is sharpened to the dictum of psychosynthesis: "I remember, therefore, I am!" Here is the clue: If the memory of "I" is loosened or somewhat erased, then the personality of "I" is changed. Consciousness is the most recent memory, which is most subject to erasure and loosening. Personality transformation becomes energy pattern modification of not only scintillating consciousness, but also of recent circulating and old stored memories of childhood.

Psychosynthesis requires that the pathological programs of E_s must be *selectively erased* or differentially loosened. The author returned to electricity and developed the electrolytic technique (ELT) for such a process. ELT selectively erases by play-back just before electrolysis, the abnormal information discovered on the psychoanalytic couch and then reprograms the patient *immediately* in an "infant-like" state with information for a new personality. Psychosynthesis supports Wiener's speculation which can be realized with ELT for the goal-directed transformation of any old undesirable personality into a new desirable one.

The psychosynthesis model is practical and can be carried out by any qualified eclectic psychiatrist today. An eclectic psychiatrist is one who subscribes to all known useful empirically proven methods for the purpose of improving a personality. Psychosynthesis is the cybernetic union of the eclectic elements of psychiatry into a meaningful, communicative system. Psychosynthesis is simply cybernetic psychiatry of eclectic elements. In this way, we have unified the whole field of psychiatry as a system. The question comes naturally at this point, "How is the system linked up together, by what means?" The answer is simply, television (Tien, 1969)[(8)].

TV-CYBERNETIC SYSTEM OF PSYCHOSYNTHESIS

The various elements such as psychoanalysis, interview psychotherapy, behavior therapy, chemotherapy, community mental health programs, etc., and sub-systems of the psychosynthesis system, linked by closed-circuit television (Fig. 1, Chap. 14). Psychosynthesis is, therefore, a TV-linked cybernetic system of psychotherapy. Having thus fused the eclectic elements of modern psychiatry into a multiple TV-linked feedback system as a mental health mini-community, the following diagram will illustrate[(7)]. (Fig. 1, Chap. 14)

The patient contacts this TV-Cybernetic system by interview as in classical approach. And then he will go through interviews with the family members and couch analysis, which are also linked by television to record the progress of family therapy, drug therapy, electrolytic therapy, group therapy, super-group therapy, through which he again returns to the community as a stable functioning personality. His consciousness is then programmed and maintained by all the sub-systems. It is important to point out here that all existing theories of psychotherapy depend on the addition of information, whereas psychosynthesis blends in an important significant element, alluded to by Wiener, namely the subtraction of information as differential selective erasure of information. The subtraction of information from memory and consciousness speeds up the process of integration and stabilization of the new personality.

THE TRANSFORMATION OF PERSONALITY

Hithertofore, all the classical theories of psychoanalysis and psychotherapies and psychosynthesis have one thing in common, the adding of information into the brain in order to modify memories and hopefully to transform personality. However, the psychosynthesis model demands the *subtraction* of information from the brain as well the *addition* of information as used by the classical theories of all schools of thought. If personality is a time-series of patterns, then its transformation may be achieved by adding and also by taking away information. Addition of information may be accomplished by verbal or visual input; and subtraction of information may be effected by electrical or perhaps by chemical means. It is now clear that the transformation of personality is the differential modification of energy patterns. Energy patterns may be modified verbally, visually, chemically, or electrically, or any combinations thereof, without contradictions of the contending schools of psychiatric thought, whether organic or psychogenic in formulation.

The most effective and economical way of adding information to the personality growth and development is the use of videotherapy frequently and intensely with both feedback and feedforward of significant videosessions.

We have developed a library of several thousand videotapes of such significant and relevant information for the purpose of personality enhancement and transformation.

SUMMARY

Psychosynthesis is a TV-linked cybernetic system of psychotherapy. Psychosynthesis is a unified theory of energy transformation. In theory, psychosynthesis is based on the unified concept of personality transformation of brain-mind equivalence in the same sense of the mass-energy equation of Einstein. In practice, psychosynthesis is the cybernetic union of psychoanalysis of Freud, behavior therapy of Pavlov, electrotherapy of Cerletti and information theory of Shannon linked by modern television technology as a cybernetic process of Wiener. And so, the psychosynthesis model unites all schools of psychoanalysis, psychotherapy and electrochemotherapy into a cybernetic system with a goal-orientation of the individual seeking its identity in the community.

With the psychosynthesis model, the author has fused the eclectic elements of modern psychiatry into a multiple TV-linked feedback system, composed of interview psychotherapy, couch analysis, family therapy, drug therapy, electrolytic therapy (ELT), group therapy, and super-group therapy as on-going dynamics of mental health mini-community of fifty families. In this system, the psychiatrist is also being transformed into a family psychiatrist, who could potentially work simultaneously with 100 families or more as the psychosynthesist of a mental health mini-community.

13. A TV-CYBERNETIC HOLOGRAM MODEL

For some years now this property of holograms has attracted the interest of neurophysiologists who were puzzled by the difficulty of locating the 'engram' in the human or animal memory. As is well known, especially since the famous experiments of Lashley, large parts of the brain can be destroyed without wiping out a learned pattern of behavior. This has led to speculation that the brain may contain a holographic mechanism.

—*D. Gabor,*[4] *1969*

Psychosynthesis[(8)] models man's brain as random-dot hologram. The human brain is a cybernetic synthesis of random bits of genetic information to maintain life through mutation, death, and feedback in the Darwinian evolution. And the human mind is a cybernetic synthesis of random dots of cultural information. An adaptable personality is the psychosynthesis of evolution and culture. An adaptable personality survives and an unadaptable one dies. The psychosynthesis model suggests that evolution continues in the human cortex, and the mutation of an unadaptable personality can be achieved without death. This is based on the Theory of Psychosphere.

THE THEORY OF PSYCHOSPHERE: 21 POSTULATES

This is a tentative outline of the theory of psychosphere, viewing the cerebral cortex of man as a 4-dimensional random dot hologram, based on 21 postulates.

Postulates 1–7: Psychosphere and Hologram

P-1. The psychosphere is man's total functioning cerebral cortex as a hologram.

P-2. Man's present psychosphere has evolved for at least two million years as an active self-organizing mirror system of the Einsteinian Universe.

P-3. The psychosphere can be likened to a 3-dimensional hemispherical holographic television screen with *memory* as the fourth dimension of time.

P-4. The cerebral cortex is the psychic screen of memory interplay of genetic and cultural information with current active sensorimotor input into the hologram. Consciousness is the activity of the psychosphere.

P-5. This holographic psychosphere continues to model itself after the Universe, evolving from the animistic to the religious, from the metaphysical to the scientific, and from the Newtonian to the Einsteinian, and so on.

P-6. The nature of the memory process is holographic, that is to say, in a small time-space, a fragment of the psychosphere can reproduce or reconstruct the whole pattern of a personality. In other words, the memory of a given personality is encoded in every portion of the psychosphere. And many personalities can coexist in the same holographic system.

P-7. The psychosphere is a cybernetic self-creating system of synchronized neural activities in the sense of a time-series of Wiener.

Postulates 8–14: Time-Series and Transformation

P-8. The synchronized activity of the psychosphere is a periodical time-series.

P-9. The stability of a personality depends on the psychological inertia of the time-series. The psychological inertia is a function of synchrony of the neuronal activities.

P-10. The psychosphere has psychological inertia, a complex derivative of the inertia of physics.

P-11. The psychosphere begins as a relatively random state but develops into a synchronized pattern of self-creation by duplication.

P-12. The psychosphere recreates its own pattern.

P-13. The psychosphere is a pattern-transforming system. Paradoxically, it transforms itself from one pattern into another as it attempts to maintain stability and identity.

P-14. The psychosphere cannot duplicate itself identically, hence its growth, development and transformation.

Postulates 15–21: Consciousness and Personality

P-15. Consciousness is a single frame of a time-series or megacycle.

P-16. A given personality can be viewed as a time-series.

P-17. Consciousness is a special expression of synchronized memory-in-motion.

P-18. The conscious pattern of the psychosphere is a function of memory. This is to say that the personality of "I" is a product of memory-in-motion, which can be expressed in a dictum: "I remember, therefore I am."

P-19. The psychosphere of a newborn child is a relatively random hologram to be co-programmed by genetic and cultural information from the external world, namely from the time-series of the mother and the father.

P-20. The name of the pattern serves as an organizer of the personality (e.g., If you name your boy "John," the name John serves as the organizer of his identity of "I". And "I" attains consciousness by the motion of self-creation in the psychosphere).

P-21. Consciousness is motion.

THE PSYCHOSYNTHESIS MODEL[8]

The model is represented by:

$$E_i(E_s) \rightarrow E_o$$

E_i = energy input pattern. E_s = stored energy pattern of genetics and experience. E_o = output energy pattern, i.e., behavior. The human cortex can be seen, then, as a random dot 3-dimensional matrix or stereogram with a cybernetic capacity for self-creation. The moment-to-moment scintillating pattern of this hologram is consciousness, which recreates itself over time.

DEFINITION OF PERSONALITY

Personality is a time-series of scintillating frames of consciousness, which uses information from inheritance, memory and environment to transform the random dot hologram of the cortex to an orderly, adaptable and responsible self-creating pattern of personality. To transform an unadaptable personality to an adaptable one, we need a psychological mutation or psychomutation.

PSYCHOMUTATION AND ELECTROLYTIC THERAPY (ELT)[9]

This psychomutation can be produced by the electrolytic process of electrolytic therapy (ELT). The ELT is used to cause all neurons to discharge simultaneously, so that the unadaptable pattern of personality is being reset momentarily to a zero-matrix. And before the neurons return to the previous pattern of activity, it must pass through a random dot state, similar to that of an infant. Before the random dots scintillate back to the fixed pattern of the previous personality, information is being immediately reprogrammed into the brain for a new mutant pattern. Based on the above model, the author has constructed a TV-cybernetic system in which an unadaptable, fixed personality pattern may be given a psychomutation similar to the organic mutation for the transformation and synthesis of a more adaptable personality for survival in the modern world.

THE TV-CYBERNETIC SYSTEM

From 1962 to 1967, the author evolved gradually the theory of psychosynthesis and the technique of transmutation of a less adaptable personality to a more adaptable one out of his practical experiences with the ELT. From 1967 to date, using the mass-energy equivalence concept of Einstein, the author unified the Darwinian concept of evolution, the psychoanalysis of Freud,[2] the conditioning theory of Pavlov,[6] the information theory of Shannon(2) to build a practical cybernetic system of Wiener[10], utilizing the modern television technology and electrochemotherapy for the transmutation of less adaptable personalities to more adaptable ones in the community. As a result, the author has formed a TV-linked mental health minicommunity of almost a hundred families, who have achieved eighty-five percent results in transforming less adaptable personalities to more adaptable ones, with relative mental stability, productivity and happiness in the complex and conflicting ecological sub-systems of the modern world.

DISCUSSION

The theory of psychosphere is based on the suggestion of the holographic nature of storing information in the brain. This idea has been noted by the originator of holography, Gabor[3], Longuet-Higgins and others. Brown[1] stated, "Indeed, it has been suggested that the human brain may store information in this way. It has always been a great puzzle how the vast amount of information stored in the human memory can be contained in the volume of the brain, and the holographic method might well be the way it is done." However, Gabor[3,4] commented that, "For my part I am inclined to believe that there exists an abstract, mathematical similarity, but I am rather skeptical regarding the existence of waves or even of tuned resonators in the brain."

The author agrees with Gabor to view the psychosphere with holographoid properties. A holographic psychosphere explains better the electrolytic process. Otherwise, how can global application of electrical current produce selective erasure of memory? We are taking advantage of the small differential between memory-in-motion and memory-at-rest. We are deliberately trying to produce memory-in-motion in order to produce lysis and erasure. Commenting on the electrolytic therapy of psychosynthesis, Gabor[5] stated, "I have hardly ever read anything as fascinating and exciting as your short note on psychosynthesis. What courage you psychiatrists have, to interfere so thoroughly with a personality!" and "Speaking scientifically, and not emotionally, the fact that memories revealed in the 'prelytic' step are selectively erased in the 'lytic' step will be, when proved, a most valuable cue for the theory of memory, conscious and unconscious."

The theory of psychosphere, the model of psychosynthesis, is still in the early stage of development. The use of the concept of holographic principle is an attempt to explain the nature of the electrolytic process, memory loosening, erasure, and personality transformation. Indeed, the combination of psychological theories with engineering principles may give us a deeper understanding of the human mind and increase our ability to predict, influence, and improve behavior in directing our evolution.

A TV-cybernetic system provides a model in the use of videotechniques for information storage (memory) and feedback in the programming of a personality. If the cerebral cortex functions as an active information organizer and cybernetic memory system, then the use of videosystem will help to provide a technique of the optimal way of information retrieval and utilization in personality function.

Videotherapy is now based on this working hypothesis that the brain functions like a biohologram, which needs constant input information and data for re-creation, reorganization, and retransformation in dealing with the biodynamic universe of people and things.

Videosynthesis is the application of the TV-cybernetic system theory in actual practice with simultaneous feedback and feedforward of information such that the therapeutic videosystem is also a dynamic one in relationship not only to the individual patient, his family, but also the other families in the community, with the family psychiatrist serving as a synthesist for mental health of the community.

REFERENCES

1. Ronald Brown, *Lasers: Tools of Modern Technology* (New York: Doubleday Science Series, 1968) p. 171.
2. S. Freud, *The Basic Writings of Sigmund Freud* (New York: Random House, 1938).
3. D. Gabor, "Associative Holographic Memories." *Journal of Research and Development* 13: 1969.
4. D. Gabor, "Improved Holographic Model of Temporal Recall." *Nature* 217: 5135.
5. D. Gabor, Personal Communication, December 16, 1968.

6. I. P. Pavlov, *Experimental Psychology and Other Essays* (New York: Philosophical Library, 1957).
7. C. E. Shannon, "Communications in the Presence of Noise." *Proc. IRE* (January 1949).
8. H. C. Tien, "Pattern Recognition and Psychosynthesis." *Am. J. Psychother.* 23:1.
9. H. C. Tien, "Theory and Basic Techniques of Psychosynthesis." *Transactions* 2: 1970.
10. N. Wiener, *Cybernetics* (New York: MIT Press, John Wiley, 1966).

14. THE PERF MODEL OF VIDEOSYNTHESIS

The PERF Model is being developed as a practical and theoretical guide in the practice of psychiatry and psychosomatic medicine, using TV-cybernetic subsystems for the automation of clinical diagnosis and therapy. The PERF Model (acronym for Program-Erase-Reprogram-Feedback) is a psychocybernetic model for the purpose of communication and the development of higher personalities. The model can be adapted not only to the general practice of medicine but to any person-to-person relationship, where the communication and feedback are essential in the development of a higher system, whether in the individual, in the family, or in the world community.

The PERF Model of psychosynthesis has given rise to the science of videology-television-cybernetic[1] system of medical practice in diagnosis and therapy. The science of videology as developed out of the PERF Model of psychosynthesis may solve national health manpower problems and, at the same time, it has the potential of equalizing rural-urban imbalance of medical service throughout the world. Let us look at the historical origin of the PERF Model in order to appreciate its practical applications and theoretical implications, based on a TV-cybernetic model of human brain.

HISTORICAL DEVELOPMENT

In 1960, the author[2] as a practicing psychiatrist began the search for a therapeutic model other than psychoanalysis. Even though some of the Freudian concepts are true, the reality of medical manpower problems and economics also play a crucial part in medical service. It is common knowledge that many patients suffer from emotional and mental disorders even though they present physical symptoms. Even if all physicians were to become psychoanalysts, there would not be enough psychoanalysts to satisfy such a manpower need and to help the public. In 1969, the author[3] first presented the TV-cybernetic hologram model of human brain at the First International Congress of Cybernetics, London, England, and also at the American Psychiatric Association annual meeting in Florida. The model can be represented as follows:

$$E_i(E_s) \rightarrow E_o.$$

E_i = input energy pattern; E_s = stored energy pattern of information and system function of the human cortex; E_o = output energy pattern, i.e., behavior. The human cortex can be seen then as a three-dimensional holographic stereogram with a cybernetic capacity for self-creation. The moment-to-moment scintillating pattern is defined as consciousness. With this model, the author guided the development of electrolytic therapy (ELT) from classical electroconvulsive therapy (ECT) for the purpose of analysis of a personality (Program), immediately followed by the electrolytic process of memory modification (Erase), so that a higher personality can be constructed with new information (Reprogram) with a personality blueprint so that gradually the person can reach his goal through constant correction (Feedback).

This is the PERF Model (acronym for Program-Erase-Reprogram-Feedback). This theoretic possibility was clearly noted by Wiener[(4)] in *Cybernetics*: "There is therefore nothing surprising in considering the functional mental disorders as fundamentally diseases of memory, of the circulating information kept by the brain in the active state." As a practicing psychiatrist, the author took special interest in this remark that functional mental disorders, such as psychosomatic disorders, neurosis, and psychosis, are due to disturbing scintillating information. The author therefore gradually advanced psychoanalysis into psychosynthesis with electrolytic therapy (ELT) based on the PERF Model. The circulating memory of a personality can be programmed, erased, reprogrammed, and fed back into the cortex in order to maintain stability. With the modern electronic technology of television and electrolyzer (i.e., memory-erasing electro-stimulator, Reiter SOS), the author has been successful in unifying electrotherapy with psychotherapy, as predicted by Wiener. This is electrolytic therapy (ELT) based on the PERF Model (Figs. 1–4).

Guided by the basic PERF Model, the author began to develop a medical clinic using extensive TV-cybernetic subsystems, simulating the human cortical activity. As a result, the author can duplicate one physician into two or three or more. This certainly reflects what Marshall McLuhan[(5)] had observed, that, "The effect of television as the most recent and spectacular electrical extension of the central nervous system is hard to grasp for various reasons." The author would like to point out that the TV-cybernetic system is not only an extension of our brain but also can duplicate our entire personality, especially our professional personality. The author derived a secondary model for general psychotherapy. This is the POPF Model.

Fig. 1. POPF Model: Physician makes Professional contact.

Fig. 2. Patient watches same physician on television.

Fig. 3. Physician makes second Professional contact.

Fig. 4. Feedback from doctor completes loop in POPF Model.

THE POPF MODEL

The POPF Model (acronym for Professional contact-On television-Professional contact-Feedback) is a direct application of the PERF Model. The POPF Model can be easily grasped as a TV-cybernetic loop (Figs. 5–8).

With the POPF Model, the doctor-patient relationship can be amplified by using the TV-taped personality of the professional. In brief, a physician can see his patient in person at the beginning of a TV-cybernetic session in order to find out what the chief complaint or the medical problem is, and then the patient is introduced to the duplicate physician on television, which is a pre-recorded treatment program appropriate to the patient's problem. While medical information is being actively transferred from the TV-tape to the patient's cortex, the physician leaves the first patient to service another patient and so on. At the end of the TV-taped program, the physician returns to close the loop with personal contact by discussing further with the patient. At this point of exit, the patient is given medication, tests, or future appointment in order to maintain the therapeutic loop. This simple POPF Model has proven a most powerful technique in amplifying medical television procedures which the author calls telefusion techniques.

BASIC TELEFUSION TECHNIQUES

The telefusion techniques can be classified into three types: 1. Primary telefusion technique; 2. Ordinary telefusion technique; and 3. Booster telefusion technique.

1. Primary telefusion technique. The physician and the patient are simultaneously TV-taped and also TV-linked with immediate feedback through television, with each in a different room. This, of course, enhances the privacy of the patient (for example, especially for couch therapy or hypnotherapy). The primary telefusion technique creates a master TV-tape, which the physician can use with other patients, even in as complex a system as psychoanalysis and psychosynthesis. Here, by going away from classical procedure, we have returned to it through technology and by making this available on a mass scale through the ordinary telefusion technique.

2. Ordinary telefusion technique. Here the physician has an opportunity to offer therapy on TV-tape with his personal physical contact at the beginning and at the end of the session according to the POPF Model. At the end of the ordinary telefusion session, the physician appears to deal with any immediate and urgent problems, especially in preventive procedures, if the patient feels the physician's presence on television was not sufficient.

3. Booster telefusion technique. This is similar to the ordinary telefusion technique except there is only delayed feedback. For example, the

physician could write a prescription of twelve weekly sessions for parents to learn about child discipline. The parents will see the physician on television alone for such a series and then at the end of three months the parents and the physician would get together for delayed feedback and discussion as to the course of the child's development. The power of booster telefusion technique becomes evident when the physician can analyze himself with telefusion couch therapy.

We can here grasp that television can duplicate the personality of the physician, thus amplifying his capacity to function as a physician for many families at once, depending on their needs. The telefusion techniques are not only effective in psychiatry to the extent we can now do psychoanalysis and psychosynthesis on a massive scale, but we can offer such service to families from all walks of life and all economic classes. In fact, any physician can apply the same POPF Model to the general practice of medicine. In this futuristic practice, we are beginning to see the possibility of an ever-expanding TV-cybernetic medical system. In short, the PERF Model has given rise to a new medical specialty: videology.

THE SCIENCE OF VIDEOLOGY

The science of videology may be defined as the technology and art of medical television. Videology is analogous to radiology in that we employ the use of a TV camera and TV-tapes for the examination, diagnosis and therapy in curing diseases or preventing disorders, physical or mental, very much like we employ X-ray machine and film in the practice of radiology. Videology is a by-product of psychosynthesis, which is a fusion of various ideas and techniques in psychiatry, where manpower shortage has been the most acute. In the attempt to seek a solution to the medical manpower shortage in psychiatry, the author has also found a solution for general practice of medicine.

In fact, medical history can be automated with a TV-cybernetic subsystem. Based on the POPF Model, the physician comes in the consultation room to greet the patient in person. After obtaining the chief complaint, he will then present himself on television so that the patient can answer his medical queries by writing down the information on prepared medical history sheets. If the patient is unable to write, which occasionally happens, then a TV programmer will help the patient to write down the replies on the history sheet. From the answers the patient writes on the history sheet, it is possible to obtain a completed medical history in every detail from past allergy history to current problems, surgical procedures, etc., to whatever minute details are desired. Special TV-tapes can be prepared for gastrointestinal, cardiovascular, genitourinary histories, and so on, as detailed as a specialist would make them. Since the TV-tape can be erased and reprogrammed, a physician can develop a series of tapes for various diagnostic categories or sub-specialties using the PERF Model. Furthermore, with the television monitor as a duplicate of the physician's professional personality, he is now able to explain at length medical, or surgical procedures, diagnostic tests, prognostic pronouncements, and whatever treatment programs the patient should have. The physician could patiently explain diets, exercise, even proper admonitions as the occasion demands. These TV-taped programs can be produced without spending extra time by using the telefusion techniques the author has described above. For example, the physician can explain to a diabetic patient the nature of diabetes and the proposed treatment of insulin and diet regimen from a different room and the physician's explanation is separately taped, while the patient's voice or appearance is being recorded on another tape. This physician's tape can then be used for another patient.

As we can see, this in no way depersonalizes medical service. In fact, we are using medical television techniques to increase personal contact. Whatever time can be saved by the TV personality of the physician can be used for more important personal contact on some unique problem that often arises in the course of medical treatment of any given patient. Videology will become a futuristic practice of medicine in which personal touch is not reduced but amplified. To anticipate the growth of videology is to understand that the human brain is more attuned to the television as the brain may be similarly structured. If the human personality in sickness and disease is an accumulation of misinformation, then transformation of information is the most effective active way to provide a medical service. Videology will become an integral part of the practice of medicine for the scientific and economic advantages are so obvious when the training of a physician anywhere in the world is a time-consuming and costly function. This leads us to discussion of the implication of the PERF Model and videology in the world health manpower question.

THE PERF MODEL AND THE WORLD HEALTH SYSTEM

Based on the PERF Model and the new techniques of videology, we may have an answer to the worldwide shortage of health and medical manpower, not only in the industrialized countries like the United States, the Soviet Union, and Great Britain and other countries, but also for nations in Asia, Africa, and South America, wherever people need medical service. Even a barefoot doctor in the People's Republic of China could amplify his therapeutic armamentarium by simply having a portable television system containing tapes from all areas to service the people whether they live in urban centers or in rural areas.

However, without a cybernetic system like the PERF Model, television can only be used very much like commercial television without appreciating the basic nature of the central nervous system. In the sense of Norbert Wiener, man is a teleological animal and as such we need continuous feedback. When we model the human brain as the TV-cybernetic system, then we learn to use television cybernetically.

It is possible to develop a center of medical TV tapes on every conceivable kind of medical and surgical procedure or public health measure, old or new. For example, in the 1970's, acupuncture anesthesia[6] as developed out of the People's Republic of China has become prominent. The TV-taped techniques can be more easily taught anywhere in the world by a series of programmed procedures, from simple use of acupuncture needling in the treatment of toothache and sore throat, hypertension, ulcers, to the most modern acupuncture anesthesia with electronic pulsers. And thus, we can begin to TV-link the major urban hospitals and medical centers to the rural and countryside, and via the TV-tapes the rural problems and medical needs and difficult diagnostic problems can also be TV-taped and sent into the cities for careful examination and analysis and recommendations. The PERF Model can be easily adapted not only to the general practice of medicine but also to any person-to-person relationship where communication and feedback are essential in the development of a higher system of medical service, whether for the individual, the family, or for the world community.

DISCUSSION

At this point, one may raise the question, why is the PERF Model so effective? The author suggests that it is because it is based on the theory of psychosphere, which suggests that the human brain functions like an active TV-cybernetic hologram. The author[7] formulated a TV-cybernetic hologram model of the human brain, based on twenty-one postulates. The first four postulates are as follows:

P-1. The psychosphere is man's total functioning cerebral cortex as a hologram.

P-2. Man's present psychosphere has evolved for at least two million years as an active, self-organizing neural system of the Einsteinian universe.

P-3. The psychosphere can be likened to a three-dimensional hemispherical holographic television screen, with memory as the fourth dimension of time.

P-4. The cerebral cortex is a psychic screen of memory interplay of genetic and cultural information with current active sensory-motor input into the hologram. Consciousness is scintillating activity of the circulating memory.

A person in health or disease depends on the circulating memory in order to maintain stability as suggested by Wiener. The theory of TV-cybernetic model of human brain also explains how some personality patterns become unsuccessful and severe depression ensues. Depression and suicide are major causes of morbidity and death in many countries. The PERF Model can also guide the psychiatrist in administering electrolytic therapy in transforming a depressed personality to a happy one without suicide. Wiener[4] has so graphically pointed out, "Thus, what started as a relatively trivial and accidental reversal of stability may build itself up into a process totally destructive to the ordinary mental life."

He offered the analogy in the machine, "How do we deal with these accidents in the use of the machine? The first thing which we try is to clear the machine of all information, in the hope that when it starts again with different data the difficulty may not recur." The author has applied cybernetic principles to psychiatry and medicine with positive results in creating the PERF Model and TV-cybernetic model of human brain.

Interestingly enough, the TV-cybernetic model is so useful that with it the author[(8)] was able to explain the phenomenon of acupuncture anesthesia, as discovered by the Chinese revolutionary medical workers. This type of anesthesia may be explained as neurogenic interference of pain signals on the cortex (screen) by peripheral electrical stimulation, similar to the "snowy effect" of television when extraneous radio frequencies can obscure the TV signal on the TV screen (cortex). However, it is beyond the scope of this chapter to discuss the theory of acupuncture anesthesia.

A good theoretical model has many predictive powers, with which we can also influence and improve our lives in general. It is out of the synthesis of engineering concepts and psychiatric theories that the author evolved the PERF Model, where we can deliberately program, erase, reprogram and feed back information in order to maintain the highest goal of life, that is, life itself.

This also points out the possibility that one of the major scourges of mankind is war and perhaps we will have the technological know-how to prevent war through the use of the PERF Model.

REFERENCES

1. H. C. Tien, "The Science of Videology: TV-Monitored Evaluation Forms." *World Journal of Psychosynthesis* 4: 30–42, 1972.
2. H. C. Tien, "Psychosynthesis: A Cybernetic System of Psychotherapy" (Sixth Annual Rocky Mountain Bioengineering Symposium, Laramie, Wyoming, 1969).
3. H. C. Tien, "Psychosynthesis: TV-Linked Cybernetic System Psychotherapy" in *Progress of Cybernetics,* ed. J. Rose, Vol 1 (London: Gordon and Breach Science Publishers, Ltd., 1970).
4. N. Wiener, *Cybernetics* (New York: MIT Press and John Wiley, 1961).
5. M. McLuhan, *Understanding Media: The Extensions of Man*, 2nd ed. (New York: Signet Books, 1964).
6. Acupuncture Anesthesia (Peking, China: Foreign Language Press, 1972).
7. H. C. Tien, "Psychosynthesis: A TV-Cybernetic Hologram Model," in *The Nervous System and Electric Currents*, Vol. 2 (New York: Plenum Press, 1971).
8. H. C. Tien, "Acupuncture Anesthesia: Neurogenic Interference Theory. On the First Case of Chinese Acupuncture Anesthesia." *World Journal of Psychosynthesis* 4: 36–41.

15. TV-LINKED FAMILY PRACTICE OF MEDICINE

As general practitioners are evolving into family physicians, general psychiatrists are emerging as family psychiatrists.

Family physicians and family psychiatrists have a great deal in common. The fact that more than fifty percent of the patients who visit family physicians need attention in the emotional and mental areas speaks for itself. The family practice approach to providing medical care may hold a key to resolving the health manpower problem. The family psychiatrist can learn from the family physician, and the family physician can learn from the family psychiatrist.

Modern Techniques Are Adaptable. The family psychiatrist can show how modern psychiatry is practiced and the techniques and methods may be useful and adaptable to the practice of family physicians. Some common problems are time limitation, waiting room complaints, and inadequate explanation of medical diagnoses, prognosis, and treatments to patients and their families. The eclectic psychiatrist is striving to resolve some of the problems and meet these needs somewhat. However, the family psychiatrist using modern television technology is about to make significant contributions in the area of family practice. For example, at Michigan Institute of Psychosynthesis, a solo family psychiatrist is able to serve 200 patient visits in a forty-hour week.

How is this done?

By television technology. The TV-linked family practice of psychiatry probably has direct application for family practice of other medical fields, especially general practice.

Whenever we have a sudden change without preparation, we are going to meet, probably, unexpected situations. This requires time. Most of us are not able to cope with unexpected situations because of lack of time. But if we can do this or do that in our mind or in discussion, rather than actually doing it, then we can predict what can happen. If anything is to be said about what we have learned from psychiatrists, it is their willingness to take time to sit down and listen. You know, most of us don't have enough time in communicating with people. The family psychiatrist must and the family physician should.

However, the family physician is a very busy and overworked man or woman. Very often in a medical meeting, everyone will be sitting down to a serious discussion, or dinner, and suddenly they will hear, "beep!" You know—one of the doctors has one of those radios in his pocket in order to keep the continuity of communication. I think communication is central to providing medical service. The capacity to hear, to see, to feel is to keep in touch with the world. How does the family physician keep up the maximum communication with the families he takes care of? And what do you do when the doctor is not in the office? As you know, one of the traditional jokes about the psychoanalyst is when he goes on vacation in Florida and lies down on the beach to relax. Suddenly, he finds one of his couch patients lying down next to him, following him on the same trip, because she wants to maintain continuity of therapy. How deep is this problem?

Family Practice and Continuity. Human beings need other human beings. It begins very early with children. The children do not let the parents out of their sight, and the wise parents usually make the proper provisions for the children. It is true for the family physician. Any family doctor who has done a good job in the community knows that he must provide coverage in his absence. Family practice can have more continuity than we realize. Families and patients are always saying that the American Medical Association should provide more doctors. We cannot provide the quality of service if there are not enough of us. We have to have time to give the quality of service. Are we able to maintain the quality of communication and availability of service? As a family psychiatrist, listening to the families, I am constantly learning how we can do this. So when a patient mentioned about the usefulness of explaining in detail and at length by the use of a TV-booster session, we begin to provide extra medical service. The doctor does not need to be there in person, but on tape his TV-personality is explaining medical matters on closed-circuit television, such as family planning, birth control pills, diet, child raising, etc.

Better medical service should not be dependent on one individual, i.e., the physician, but rather on the community at large. Mental health or physical health should be based on the community, on the families, not on the one professional person alone. I think TV-linked family practice could go far, if we can show how TV-linked family practice works. It does not mean we have to have acceptability for everybody every time we use it. And that comes later. The community must be acquainted with it.

Let TV Do the Talking. Baby-sitting with patients isn't necessarily good medicine. If the family doctor is tied talking to one family, then nobody becomes productive. Yet the family needs the information. Solution? Let the TV do the talking. Let the TV-personality of the doctor do the explaining on TV. If you are tied to the situation without other ways of maintaining continuity, then you are not able to be more productive. I think a hard-working, solo family doctor has probably done well without TV technology, but the time has come in which he can make provisions for earlier and less expensive means of communication. That is the key to maintaining quantity, quality, and stability of medical care for the average family.

And it is the wise family physician who could think consciously ahead of time of what provisions have been made for the families in his care to learn to accept better medical care with less cost, by the use of TV technology appropriately in explaining medical and surgical procedures, diagnostic tests, prognostic pronouncements, encouragements, and therapeutic programs. It is not by accident when the patient fails to take this medicine faithfully or to follow medical procedures wholeheartedly. It is too often because the physician has not taken the time to explain the diagnosis or therapy in depth during the treatment. It is not that the physician does not *wish* to take time, but he does not *have* the time. TV technology in the hands of the family psychiatrist has overcome this, and I believe it is applicable to the family physician and other medical specialists as well, if judiciously used.

TV Solves Manpower Problems. The psychiatrist, like the general practitioner, is moving gradually from solo practice to family practice. And the family psychiatrist is becoming a psychosynthesist, who uses TV technology extensively to amplify and to improve the quality and quantity of the therapeutic hour. And this will give us a solution to the manpower problem of medical service, and, of course, gradually families are learning that other doctors could learn to do the same. TV-linked practice will grow from Michigan to New York and California, as the families learn the advantages of TV-linked and TV-booster medical appointments in reducing costs and frustrations.

We have shown this work at the American Psychiatric Association the last three years, on TV-linked family practice of psychiatry. We think the TV-linked system is applicable to other medical specialties, especially the family practice of medicine.

To look ahead, we would like to make a few futuristic statements that perhaps would be helpful for different types of medical practice as well as for family physicians to improve medical service.

Consultants on TV Tape. Let us look from the family physician's point of view. He could have a series of TV-tapes on the various important consultants to whom he refers his patients, such as cardiologist, diabetic specialist, pediatrician, obstetrician, allergist and, of course, the family psychiatrist. Instead of just giving the family a consultant's name, which is very abstract and impersonal, the consultant can be introduced "live" on television, as the family physician shows his confidence in the specialist to whom he is about to refer his patient, giving the specialist a chance to explain on television his approaches, procedures, and possible costs involved, even where his office is. In fact, the family may wish to see two or three alternative consultants on television, in order to determine which one he, she, or the child may relate best to, as the case may be.

Speaking of the child, we can see how important TV-linked sessions would be to introduce a child, who is about to be hospitalized, to the head nurse, the lab technician, the hospital procedures, and the total hospital environment. This could be done on television without the child having to make special trips to the hospital, even though that can still be done, and additional information is available in an instant at the touch of a dial. For example, a family physician who is about to refer the mother and father to have the child's tonsils removed can introduce the ENT specialist on TV and show the child at the same time some parts of the procedure at the hospital to alleviate his fears and those of his parents.

Certainly, in the futuristic practice of TV-linked family medicine the case studies, diagnoses, and the progress report can all be TV-taped. In fact, the family physician could have the whole family case transferred via TV tapes. As we are discussing diagnostic, therapeutic, or various special procedures, we are beginning to see it is an ever-expanding possibility in the development of the TV-linked practice of family medicine and psychiatry. It may be before 1984 that home televisions would become TV-phones for futuristic medicine!

Stability and Familiarity. At this point, I would like to add a word of humanistic concern. We are not saying we should necessarily show indiscriminately and arbitrarily procedures and techniques as on shows such as Medical Center or Dr. Kildare, but rather Dr. John Smith, the family physician. The television-linked appointments are not less personal, but more. The television personality on the monitor is not *any* physician, *any* consultant, but *the* family physician and *the* family consultant referred by the family physician. This is our psychiatric insight, namely, to introduce stability by having familiarity. This is the key of what we mean by transference. Transference (or unique rapport) is based on the conveying of a significant quantity of relevant information from the family physician to the family members, and vice versa. This I believe can be done economically, effectively and efficiently by TV-booster sessions.

CONCLUSION

To have good rapport, we must develop optimum communication between the professionals and the families, not only for curing illnesses, but for the prevention of diseases, and perhaps even pointing toward a creative cooperation for the physical and mental well-being of all families in our own community and the world community.

In the world community, the TV-linked practice may be even more important than we realize in the developing and underdeveloped countries of Asia, South America, and Africa, where the modern medical manpower shortage is even more severe than in the industrialized nations.

16. TV-LINKED GROUP THERAPY REVISITED

FAMILY MEDICINE AND VIDEOLOGISTS

SYNTHESIST: Good afternoon, everybody. I think most of you remember Dr. Hatton. He practices family medicine. Several months ago Joyce asked who would be the first doctor to become a videologist. If I am the first one, Dr. Hatton would be the second one around here. Videology is like radiology. They use X-ray and films, and we use television and tapes. Everything we do is always emerging anew to help the families.

As the world gets a little faster, we have to keep up with the change. Dr. Hatton has tremendous background on physical illnesses and family problems, and he is going to share with us. I am going to let him take over the group as a group coordinator. Dr. Hatton, what we do is to elicit from everybody what they have on their minds. You go from person to person, family to family. They may have some tough questions for you. You will find out the wisdom of the family is infinite.

DR. HATTON: All right. Let's start over here.

EDWIN: I am Edwin. I have no problems, everything is going along fine.

ROGER: I am Roger. I don't have any problems.

SYNTHESIST: This resistance, as Freud talked about, is natural. They say, "I have no problem," but of course it is not true. Let's take Roger, for example. Would you want to tell us what you are up against now, Roger? Dr. Hatton doesn't know anything about you, but he is a fully qualified videologist. Whatever you want to bring up.

ROGER: The biggest thing I've got against me right now is my drinking problem. I don't drink very often but when I do it has fantastic repercussions. I usually wind up in a world of trouble. I am trying to do something about it so it doesn't happen again. Usually I get arrested for driving under the influence or have policemen chasing me all over town. I found since I started here that it does help.

RICHARD: I feel pretty good today. I don't really have much to offer.

DR. HATTON: Well, it takes time. We will get acquainted. I do a different type of practice than your synthesist does, but I am intensely interested in this field.

ROY: I am Roy and I have been coming here for quite some time. I have gotten a tremendous amount of knowledge from the group and booster sessions. Before I came here I was what you call a shopper. I thought I was sick all the time and I didn't know why. Our family doctor said we should go see a psychiatrist.

DR. HATTON: How did you feel about that?

ROY: I knew something had to be done but I didn't know what. He took me into his office and said, "Roy, I've done everything medically that I can do for you. You need somebody with more advanced training to help you out of the situation. Would you go?" I said, "I'll go if you make the appointment for me, but get me a good one or I am not going." "Okay," he said, "I'll do that." So he got right on the phone and called and two days later I was here.

DR. HATTON: What type of problem were you having?

ROY: I was losing time from work being sick, and if I had a cold or anything it just seemed like it would build up. I was making mountains out of molehills.

DR. HATTON: Things are better now?

ROY: Now I am able to cope with it.

VALERIE: I was having the same kind of problem. I felt sick to my stomach and I was sure I had cancer and different things and the doctor told me I should see a psychiatrist and that's what I did.

LOSING TEMPER WITH CHILDREN

MARK: I'm here to get rid of the bad temper. I guess you would call it that. My wife and I were about ready to get a divorce and we decided the last thing we would do is come to a headshrinker—that's what my opinion of him was. He's helped me a lot. Really the only thing we argue about now is the children. I let them get on my nerves and I just can't stand it. I would rather work twenty-four hours a day than come home and listen to the kids. I would like to have somebody help me figure out a way not to let it bother me so much.

JANE: I have been doctoring here for three years and have been in the group most of that time.

DR. HATTON: What type of problem are you having?

JANE: My problem at that time was suicide, potential suicide. I think I have worked most of that away. There are times when it does come back in my mind, but I fight it hard. I have never attempted it. I have found coming to the synthesist and group has helped a lot to overcome some of the other problems that I have.

SYNTHESIST: We have two new members. We will let them introduce themselves.

PAT: I am Pat.

BUD: I am Bud.

SYNTHESIST: What do you hope to accomplish by coming? What do you hope to contribute?

PAT: I'd like to sit in one time and I would know better how it functions.

BUD: The same thing, I guess.

SYNTHESIST: You want to listen first before you get too deeply involved? Shall we give them a welcome? (Group applauds.)

GERI: I have been coming here since March. Before I started coming I was just getting more depressed, filled with anxiety and frustration and all kinds of terrible feelings. I didn't really know why. I have had some ups and downs since. I was in the hospital for a few days a couple weeks ago and that gave me some determination to help myself. One question I need some help on is nightmares and how they affect your everyday life. Every night I have nightmares. I am holding back from sleeping so these nightmares won't come, and that leaves me exhausted all day.

GLADYS: I have a blood pressure problem. It wants to go up and then my nerves get so bac' I can't take it. They can't tell what it is. But since I have been coming here for a couple of years, my nerves have stabilized. I have a stabilizer for my nerves, and then they can take my blood pressure again.

DR. HATTON: It is down now?

GLADYS: No, it stays right there. But they couldn't even take it before.

DOROTHY: I came for my husband's sake and right at first he needed help. He had a problem in his past that he couldn't shake and it seemed to evolve around his family. We've gone through twenty ELT treatments, and I have gone through a completely different husband—from

a very unhappy individual to someone that I think, unless he is putting on quite a good show, is pretty happy most of the time. He can take his family or leave them now. I have a question, how do you convince somebody that there is a problem when he doesn't believe it? My sister needs psychiatric help and my brother-in-law refuses to see that she has a problem, so she is not getting help. What are some of the reasons that people won't realize that there is a problem?
ALICE: My main problem is to keep myself from being depressed. I find that I get back into this problem of being unhappy and depressed, but I have been coming here for quite some time and have been helped. I have had ELT treatments, and I think I am fairly well stabilized now, although I am still taking medication.
JOYCE: I have been coming here about eight months. It was symptoms that brought me here, like hyperventilation and difficulty in breathing, depression, and anxiety. Then we have a lot of family problems. We have a son who got involved with drugs and a schizophrenia illness was brought to the surface. We have had financial problems and, well, quite a few problems, but we are learning to solve them as they come along, and we are making it.
DR. HATTON: Good. I am glad to hear that.
FRANK: My problems are the same as my wife's, but we are getting along pretty good. My problem now is trying to buy a different car than what we have. Ours is just about ready to go to pieces and I am just trying to figure out whether to put myself in debt or not, but I am going to.

HOW TO DISCIPLINE CHILDREN

DR. HATTON: Now I would like to hear some comments regarding Mark's problem. Some of you are mothers and fathers yourselves, and I would like to hear what you have done to solve this same type of problem. As a parent, I can tell you there isn't anybody who doesn't want to yell at the kids once in awhile.
RICHARD: I have three who are grown now, but I disagree with what you said. I don't think it is necessary to yell. When mine were small and had a licking coming, they got it. If they had a reward coming, they got that, too.
MARK: I believe in discipline, that's what the problem is, but I get a little bit carried away sometimes. I get mad over little things that I probably shouldn't, and I shout at the kids when I probably shouldn't, but I never spank them, very seldom. I do believe in it, but my wife doesn't believe in it, and she doesn't believe in hollering or anything at them. The major problem is my wife and I were both married before and the two boys are hers from a previous marriage, and she won't let me correct them at all. She does let me correct them, but when I do she will butt in, and I get mad when she does this because then I feel she is trying to get at me for some reason. But since I have been coming to the

doctor I have been seeing more light all the time, that it is actually me that is causing the problem. I am undecided yet whether the kids are spoiled or not. Really, I think they are spoiled, but I can never convince my wife of that.

DOROTHY: What kinds of things upset you, Mark?

MARK: My wife will go along with telling them five or six times to do something. I will not go along with that. I will tell them at the most twice. If they don't do it, then I get mad. I make them stand in a corner or something. I don't clobber them, I don't beat them, I am not that type of person. I see that every day in my job. I don't think I am that type of person, but when I let it go two times and three times then I get mad and my wife jumps in and says, "You shouldn't do this and you shouldn't do that."

DOROTHY: Do they respond when you punish them that way?

MARK: No, they don't pay attention to anything.

RICHARD: It is easy if it is the partnership that made the children in the first place. But when he married her, of course he accepted the children, too. At the same time he accepted responsibility he had to get in the cycle, in the circle. I think he is right, myself. They sound spoiled. When my kids were young, I told them once quietly, and that was sufficient. They are 18, 20, and 22 now and they have never been juvenile delinquents, they are not drunks, they don't smoke, very little swearing, they go to church. I am real happy with them.

MARK: The synthesist, though, would go along with telling them 15 or 20 times, and most people don't agree with him.

PARENTS SHOULD AGREE

JOYCE: I don't think either one is all right or all wrong. I think his wife has a tendency to overprotect them since they are hers from a previous marriage. I think both of them will have to agree on a form of discipline, one way or another. I mean, if the one doesn't believe in spankings or physical punishment, then they both should agree somehow on this. I think it is almost impossible not to lose your temper or yell at a child on certain occasions but I think your speech muscles should be controlled. You should try, you should make an effort, understand what you are disciplining the child for. Many times we become irritable over such little things, like a child talks out of turn or is supposed to take a bath and doesn't respond right away or is supposed to wash his hands before a meal, and so on and so forth. These are really very trivial. You can talk to him, explain why you wanted him to wash his hands, and if the child just doesn't abide by this, I think both parents should come to some agreement on what form of punishment would be good.

RICHARD: I don't think it is necessary to have a family conference every time one of the children is going to be corrected for not washing his hands.

JOYCE: I didn't mean each time, but I think there should be a mutual agreement, because if one wants to spank the child and the other doesn't believe in spanking, the parents are going to end up arguing between themselves.

RICHARD: Well, if he is correcting them he should do it his way, and if she is correcting them she should do it her way.

JOYCE: In the long run, this upsets the child because the parents are in duality of what form of punishment the child should receive. The child is going to be thinking, "Should I believe in my mother or should I believe in my father?"

JANE: Sometimes children like to see their mother and father have turmoil over these things. It makes them feel like they have really accomplished something.

MARK: I get real strict with the kids sometimes and my wife gets upset with me, but when I go to work, the kids hate to see me leave. I think they really go for me, but when I talk to my wife she makes me think they don't go for me at all. But I really hate to say anything when she is not here to defend herself. I don't go for that either. I do have a problem I need help with, though. The youngest one is five, and he has a habit of twisting his hair, pulling on it all the time, and no matter how many times you tell him to stop, he listens for about thirty seconds and he is right back at it, unconscious as can be, you know. I was wondering if you should speak to him about it all the time or whether you should just let it grow and hope that he gets out of it on his own.

DR. HATTON: Richard, what would you do about that?

RICHARD: I would take the shears and cut off the hair he wanted to pull on, simple as that. I don't think I did so bad, I don't have any juvenile delinquents. I guess I am bragging.

DR. HATTON: Joyce, what would you do?

JANE: Is that what you call a tic, a habit? My son doesn't do it anymore, but he used to rub other people's hair. This was quite annoying. He would do this to his brothers and sisters and of course they would become aggravated by it. At first I would become annoyed by this and speak to him, but then I realized that there are other things like that—like people who bite their fingernails or itching, moving your muscles, and so on, different forms of tics. You can repeatedly tell a person who has something like this to stop it, but it won't do any good. You have to ignore it, and eventually they will come to the realization that it doesn't look good or it becomes annoying to themselves and so they try hard to stop doing it. My son used to shake his shoulder all the time, he doesn't do that any longer, either. I have a son who stutters, and some people will laugh or tell him to stop doing it, but you just can't do this, because he doesn't have much control over it.

CRISIS IN THE COFFEE SHOP

WAYNE: I am here to tell the group about a crisis in the coffee shop. Our sales are down from what they need to be to cover our expenses

and the salaries, of employees have increased because of the increase in the minimum wage, so not all our suppliers are getting paid. We also have a premium due on insurance, so there is a total of $1000 needed to meet this crisis. We do this by buying shares of stock. Shall we go around and find out who will buy stock today? A share is $10.
(Most of the families raise their hands when asked if they support the idea of the coffee shop, but a number say they are unable to buy stock at this time.)
SYNTHESIST: I am introducing our group to Dr. Hatton and what he can do to practice what we are trying to do. As you can see, problems have been raised from child rearing to tics and the coffee shop. How do they all tie up? I will tell you, from my experience working with the families, the more you are willing to help others, at the drop of a hat, the more likely your family will grow strong. When you listen to Roger, he was describing the old personality, Bill, who spent lots of money for alcohol, getting into trouble. It is critical in life to be able to separate what is significant and what is not. How much do you think Bill spent in a year for drinks, Roger?
ROGER: That's a good question. In one year? $3500 ought to start it.
SYNTHESIST: Here you have $3500 to produce alcoholism, and when it comes to something to help what is really creative, to provide jobs for the family, for community health, for ten dollars they would say no. And this is the nature of what we would say is the more primitive personality. If you would buy a share, I will buy ten shares, Roger. It is not money at issue, really, it is the idea behind it. When you have a crisis, when people call up the doctor and say, "We need help," the doctor is here to help. We don't really worry about the financial aspects. Of course, people have to pay in order to keep things going, but when you have an emergency, you call up, we will do it. And families' Coffee Shop is related to family medicine. Dr. Hatton is not here by accident. He has a very active practice involving hundreds of families. Why do you think he is here? Because he and I see eye to eye that the families must work together. They may think, "Oh, it is your coffee shop, don't bother me." I could say, "Oh, it is your family, don't bother me, either," or "It is your child, you take care of it." But you begin to see that it is important to help other families. Suppose you close the coffee shop right away, what happens? Annette would be out of a job, and there are several members who have been doing well who wouldn't be able to work. You are really talking about how to help each other. Sometimes you have to see more than the personality who sees the $3500 in a year. That is difficult.
ROY: I was wondering how a man could drink that much.
SYNTHESIST: That is why they are alcoholics.
ROGER: The reason that I hesitated to buy shares is because I have outstanding bills, the lawyer, yours.
SYNTHESIST: Bill has never owed any doctor or lawyer as much as $3500. We know that what is spent on alcohol consumption exceeds what is spent for all the medical care in hospitals, for operations, for treatment, X-ray, blood work, everything. Can you imagine? This

amount spent on drinking alone. But they will not help the coffee shop with the amount spent on one bottle of whiskey, or two maybe. This shows the personality that is constantly in trouble.

EDWIN: Is that the right approach in the coffee shop, the monetary approach, or is there some other way to get the coffee shop back in order?

SYNTHESIST: The families have taken out $22,000 in wages from the shop.

ANNETTE: No one happened to say that the families who work there are buying stock all the time; from their pay, they are investing.

HELPING OTHER PEOPLE

SYNTHESIST: That is a very good point, Annette. People are producing ideas, you have got something going. Actually the coffee shop has created work, therapy, community mental health around here to make things possible. Some of the ideas I have learned from the families over the years I am going to take with me to the Fifth World Congress of Psychiatry in Mexico, and also to the International College of Psychosomatic Medicine.

I have no doubt the coffee shop will go on, even for those who don't vote for it. It is not difficult, say, for Dr. Hatton to buy a share. That is not the point. It is when you have the courage to help other people. When I am in Mexico to help other doctors, when Dr. Hatton will be here to help us, that is what counts. We cannot cope with all the problems alone, we have got to help each other.

Videology is part of it. We are becoming aware the world's information goes through our heads so fast, ten to 100 times as much as went through the heads of our parents. When I go to the meetings, I am taking some of the tapes with me, of the families and the coffee shop, to show how other families do it. Dr. Hatton has started to use videology in his practice, and he uses television to diagnose, treat and prevent sickness and disease. Moreover, it could be used to help prevent diseases, whatever the problems may be, whether diabetes, polio, alcoholism, childhood disturbances. Dorothy helped her husband with the problem of another alcoholic trying to break up the marriage, and now she does not hesitate three years later to help the coffee shop. It does take time, Roger, don't blame yourself. It takes time to learn what we are trying to do. It is very natural to say no. It is easy to vote with your hand, but when you vote with your pocketbook, it is something else. That means you have to have knowledge to be convinced.

Family medicine is the trend. Dr. Hatton's professional association, the American Academy of General Practice, has changed its name this very year to the American Academy of Family Practice. If the husband is sick, the doctor doesn't just take care of him, but the family, the wife, must step in. If a child gets sick, the husband and wife must both step in. If the wife has problems, the husband should come. Only in the cooperative approach do we get results. That's what the coffee shop idea is. It is when people work together. Now if Bill's marriage broke down, it is because he is only interested in drinking up the $3500 the wife wants. If the wife says, "You don't do this for me, you don't do that for me," after a few years the marriage breaks down. Now you know why. If the coffee shop goes downhill, and you had a chance to help, then at this moment you know what has happened.

The idea applies to everything, to the whole world situation. To have good health in the world, we must help other doctors. That is why I will be gone for two weeks to bring the information about videology and family medicine to other doctors. Meanwhile, Dr. Hatton will be the medical coordinator for this group during that time. Please pick his brains, he has a lot of information there. The way we end the sessions, Dr. Hatton, is to summarize all the knowledge and information, put it all together.

DR. HATTON: We discussed primarily Mark's problem this afternoon, and we had a controversy going before Richard had to leave. Basically, it is a question of cooperation between those who are involved in the situation. One must help the other. If you don't, it won't succeed. You can't be at loggerheads with one another about problems that involve members of the family who are around you. Investigate and find out what needs to be done, approach the problem more logically. It does not help to yell, but consistency is one of the main things that is lacking and consistency is the thing that would be the most beneficial. If a child is going to be yelled at, he is going to be yelled at; consistency is the thing. The father and mother should talk it over and make some overall pattern of consistent treatment of the child, so the child will know there is a unified type of front. It should involve the child himself, so the child doesn't feel he is someone who is being set aside and all of a sudden focused on for some type of reaction. Bring the child into the group and make him feel a part of it. The approach you have, in one instance, is good, because when you direct attention to the child you are immediately bringing him into the group, but it is the method by which you bring him in that is very important. It need not be one where you spend a great deal of time yelling at him, but trying to give him confidence and problem-solving abilities will enable him to be more useful to the group as a whole.

TRY TO BE CONSISTENT

The thing you have to do is work together and take issues as they present themselves, finding out the material that contributed to the problem and what you can do about it. Whether it is the coffee shop or someone who has a friend or relative who is mentally ill, or whether it is a child with problems with drugs, whatever it is, something contributed to the difficulty. It is up to the people who are a part of it to focus their attention on why and what can be done about it. Many times, the group here, due to common problems and background, will be able to do this for each other. This is the primary function of myself as a doctor. I do this day in and day out. It is the primary function of this as a family group, to be able to present some answers to the problems that can occur. You have to study the problem, find out more about the things that contribute to it. The coffee shop fits into this because as part of the family group type of response to difficulties, it presents one more source of information and one more source of available help. I don't believe that anything that is going to constitute a help should be denied.

HOW VIDEOLOGY HELPS

SYNTHESIST: Do you want to comment on videology? For instance, if we only treated one person at a time, if you only treated Roger, for example, what about all these other families?
DR. HATTON: There is no question about it. When you are able to take care of a number of people at once, as you can with television, then you have the ability to project yourself, to take care of more people and help more people at the same time. When you talk to somebody, you are helping somebody else solve their problem, too.
SYNTHESIST: Any more questions for Dr. Hatton at this point? Hopefully I may get a program of what I learned from the other doctors in Mexico. They are coming from all over the world—Hungary, Russia, China, Pakistan. We will learn what they are doing to help their families as we are trying to do here.
DR. HATTON: I am certainly going to try to help at the next two sessions.
SYNTHESIST: The families will be able to discuss more about soma, instead of psyche.
JOYCE: What is soma?
SYNTHESIST: Body. You see, I am eighty percent with psyche, twenty percent soma. Dr. Hatton is eighty percent soma, twenty percent psyche. But we are both videologists.
DR. HATTON: I will see you all next week.

17. PSYCHOSYNTHESIS, VIDEOTHERAPY, AND PSYCHIATRIC MANPOWER

Every psychiatrist must concern himself with the problems of time demands, cost and effectiveness of his therapeutic techniques and programs, regardless of his school of thought. The psychoanalyst, using couch therapy four or five times a week at fifty dollars a throw for two or three years at a time, is vulnerable to criticism. It is simply too costly. It is not available because of a shortage of psychiatric manpower, and furthermore, its effectiveness is difficult to measure. And it is too slow for many families. Marriages are breaking up, children are going haywire on drugs, and wives are threatening to commit suicide and husbands want to leave home. The families can't wait for three or four years on the couch, and bear the cost, which is often overwhelming for the average family. The problems of cost and effectiveness and time requirement have caused some psychiatrists to say that the best treatment should also be the cheapest. Robert Gibson made this observation.

We need better and faster and more economical techniques and programs in mental health. And psychoanalysis has offered the first step in modern psychiatric theory, but now we need to take a couple more practical steps to arrive at psychosynthesis.

What is psychosynthesis?

HISTORICAL NOTE

The term is not new. In 1924, Freud commented that, "I cannot imagine, however, that any new task for us is to be found in this psychosynthesis. If I were to permit myself to be honest and uncivil, I should say it is nothing but a meaningless phrase." Empty term! That's what Freud said in 1924.

Previous to that year, in 1923, curiously enough, John Baird made an announcement in the *London Times*, seeking to "hear from someone who will assist [not financially] in making" a television model. His notice went unanswered. I am sure Freud didn't pay much attention to it psychosynthesis. In 1964, and ever since, Akio Morite of Japan avenged John Baird's neglect with a plethora of video-TV models.

PSYCHOSYNTHESIST AND TELEVISION

Television equipment to the psychosynthesist is as inseparable as the X-ray machine to the radiologist and microscope to the pathologist. The psychosynthesist, equipped with television technology, modern pharmacology, and the electrolyzer (e.g., Reiter Machine), is able to amplify his therapeutic capacity many times. Psychosynthesis appears to be an answer to the problem of psychiatric manpower of today.

Today, in 1971, we *must* solve the problem of psychiatric manpower to serve the families, by the hundreds and thousands, who are faced with divorce, delinquency and drug problems, etc. If we are going to take care of people only on the couch, the Freudian way, forget it! I am convinced many personalities of Hitler's or Mussolini's type will be developed to blow up the world before we solve a fraction of our mental health problems. We cannot wait after World War III; we must prevent it!

We need preventive mental health programs now, as well as curative ones. We must offer broad-spectrum techniques, efficient treatment on mass production scale, economically and effectively, throughout the world.

A psychiatrist cannot sit behind his couch, taking care of only the urban rich or the suburban well-to-do. A psychiatrist should treat the working families as well, as the policeman sees them in the real psychodrama in the community. The psychiatrist must be a family psychiatrist and treat many families in a community setting, with continuity, accessibility, stability, and economy. Only continuity is considered by the psychoanalysts, but medical expenses and money problems often interrupt analysis even for those few motivated souls, while the vast number of families remain unserved. What can we do?

THE PSYCHOSYNTHESIS PROGRAM

The psychoanalyst could become a psychosynthesist. The psychiatrist could become a family psychiatrist. The good treatment must have support of the families in the community. The therapeutic wisdom really comes from the families, even Freud's discoveries came from the families, little by little, gleaned from what actually happened in the homes. The psychoanalyst doesn't make house calls often, but a policeman does. A city policeman could tell you a great deal about family fights, neighborhood disturbances, sexual deviations, alcoholic quarrels, broken homes, neglected children, suicide and homicide, poor mental health of many families in need of inexpensive, accessible, and effective psychiatric programs. In psychosynthesis, we attempt to put to practical use the insights of psychoanalysis and psychotherapy, and the advances of eclectic psychiatry, together with information theory, system model and electronic technology of television. Psychosynthesis is a system approach to modern psychiatry.

To comprehend the development of the psychosynthesis system, we must look at a few figures to appreciate its applicability to every psychiatrist, regardless of persuasion. Figure 1 explains the evolutionary relationship of psychoanalysis to psychosynthesis:

Table 1

	PSYCHOANALYSIS	ECLECTIC PSYCHIATRY	PSYCHOSYNTHESIS
Orientation	Individual	Variable	Family-centered
Scope	Couch therapy	Analysis; drugs, individual or family therapy; group, etc., etc.	TV-linked system of psychoanalysis and eclectic psychiatry

The orientation of psychoanalysis is individual, couch in position, and with a long-term approach. In eclectic psychiatry, the orientation is variable, including drug therapy, individual psychotherapy, group therapy, electrotherapy and many other modalities. However, they remain an eclectic bag of tricks to the practitioner without a coherent system or theory. The orientation of psychosynthesis is family-centered. When we say we are family-centered, we also unite the families into a community effort known as the Families' Coffee Shop. The coffee shop provides a community setting for all the families to interact constructively. Here psychosynthesis builds a meaningful relationship between the Freudian, Adlerian and Moreno elements of psychiatry. Psychosynthesis thus provides both intensity and depth of therapeutic effects. Figure 2 compares the number of different individuals seen in this evolutionary process:

Table 2

	PSYCHOANALYST	ECLECTIC PSYCHIATRIST	PSYCHOSYNTHESIST
No. of Different Persons Seen/Wk.	10	50	100

In the model of Freud, the psychoanalyst can see at most ten different individuals per week, if he practices classical approach. With the emergence of eclectic psychiatry and the use of drug therapy, an eclectic psychiatrist can see up to fifty different individuals per week. However, in psychosynthesis we are able to see 100 different individuals per week.

Figure 3 shows the potential number of patient visits per forty-hour week. A patient visit is defined as a person having a definite therapeutic session:

Table 3

	PSYCHOANALYST	ECLECTIC PSYCHIATRIST	PSYCHOSYNTHESIST
No. Pt. Visits/ Week	40	100	200

The figures are admittedly theoretical and estimated but the psychoanalyst to psychosynthesist amplification ratio is 1 to 10. An order of magnitude! A very hard-working psychoanalyst may have a forty-patient-visit week. An eclectic psychiatrist, seeing couples and children, combined or short sessions, may count up to 100 patient visits per week. However, the psychosynthesist is able to handle up to 200 patient visits per week, and this *does not* include the therapeutic hours in the coffee shop.

Immediately, knowledgeable professionals may ask if this magnitude of patient visits by the therapist does not dilute the therapy tremendously. Doesn't it cut down the therapeutic hour to a very brief visit in psychosynthesis? This is not so. We shall see why in Figure 4, which shows the different lengths of patient visit in hours.

Table 4

	PSYCHOANALYST	ECLECTIC PSYCHIATRIST	PSYCHOSYNTHESIST
Length of Pt. Visit	¾ hr. to 1 hr.	¼ hr. to 1 hr.	¾ hr. to 1½ hrs.

As you can see, the analytic hour is about forty-five minutes, the eclectic session may be fifteen minutes to an hour, whereas the psychosynthesis session may range from ¾ hour to 1½ hours, depending on the procedure used. So, as you can see, the therapeutic effect is not diluted in psychosynthesis. Rather, it is intensified. How is this accomplished? By television technology. Of course, this requires the construction of a closed-circuit television system. Is this costly? Yes. Might this increase the cost of the sessions? No. Figure 5 compares cost per week:

Perhaps, we can appreciate this by listening to the observation of a non-psychiatrist, Marshall McLuhan, who said: "The effect of television as the most recent and spectacular electrical extension of the central nervous system is hard to grasp for various reasons." I would like to point out that the television monitor is not only an extension of our central nervous system, but it is actually a duplicate of our personality, since we can duplicate our personality on television.

And now (in 1971), we have the technology to transform all practices of psychiatry to a system approach that has been also prophesied by Roy Grinker and Ludwig von Bertalanffy. And psychosynthesis provides the framework for change. The psychoanalysts have only *interpreted* the mind in various ways; the point, however, is to *change* it. If we are to change our patients' personalities for the better, we may have to change our own. That is to say, we must not only be psychoanalysts, eclectic clinicians, but we must also change and become psychosynthesists to answer the needs of the families in our own communities and throughout the world.

18. VIDEOLOGY AND MEDICAL ECONOMICS

The following table is a comparative study of fees charged by the classical psychoanalyst, the eclectic psychiatrist, and the videotherapist psychosynthesist for a single clinical visit. (See Table 1.)

Based on the Michigan Relative Value study, 1970–1972, the relative value measured in units may be compared as follows:

Table 1

	Classical Psychoanalyst	Eclectic Psychiatrist	Video Psychosynthesist
Relative Value/Visit	10	5 to 10	1 to 10

However interesting, this comparative chart is more revealing of the nature of economics when the services of the videologist and videotherapist are compared with that of the radiologist and radiotherapist. The technology of X-ray equipment transformed the physician into a radiologist; and the technology of TV is about to transform the eclectic psychiatrist into a videologist and videotherapist, with great economical implications in the fields of medicine and mental health. Let us study the Michigan Relative Value reports.

The units used are based on the Michigan Relative Value study, 1971–1972, which included, of course, other medical service, such as medicine, surgery, radiology, radiotherapy, and pathology. It is to be noted that in this Michigan Study, the service of psychiatry is limited and restricted when compared with medicine and surgery, and the service of radiology and radiotherapy dwarfs psychiatry to relative economical lack of value. And yet, we all realize the need for mental health service is great and a growing one in the health field. But why should a graduate physician specializing in radiology diverge so greatly from his classmate physician specializing in psychiatry after medical school? In the range of service offered? And in the efficiency, economy, and in effectiveness?

The above questions may be answered with one sentence: the radiologist uses technology extensively.

The psychiatrist who uses technology extensively would be able to offer comparable, if not greater, range and effective therapeutic programs. This can be grasped by studying the following three lists of services:

(1) The Michigan Relative Value Study of Psychiatry services;

(2) The Michigan Relative Value Study of Radiology and radiotherapy;

(3) The Michigan Institute of Psychosynthesis Relative Value study of Videology and videotherapy of family psychiatry.

LIST I

MICHIGAN RELATIVE VALUE STUDY (1971–1972)

Services	*Psychiatry*	*Value*
90800	Psychotherapy full session office-hospital-or-home verbal, drug augmented or other methods.	7
90820	Therapy convulsive electroconvulsive or drug induced, in-patient or out-patient, with or without anesthesia by treating physician.	6
90830	Therapy insulin shock, hypoglycemic, subcoma.	6
90840	Psychologic testing, psychometric and/or projective tests, with written report, given by or under supervision of physician.	5
90850	Psychotherapy in-patient care including psychotherapy and supervision of milieu team, such as occupational therapy, psychiatric nursing, etc. or conference with family, per week.	24
	Psychotherapy, individual; half session (20–25 minutes of treatment).	3.5
	Psychotherapy, individual; limited (brief) session (less than 20 minutes).	2
	Group psychotherapy, 90 minutes of treatment; group maximum of eight (8) patients—per patient.	3
	Complete psychiatric examination or consultation (if different from psychotherapy as listed above).	10

List I shows clearly the proverbial inefficiency, limitations, and the narrow-range of services offered by the psychiatrist, whose medical capacity was further limited by the practice of psychoanalysis and has been somewhat augmented by the eclectic psychiatrist using chemotherapy and electrotherapy.

List I demonstrates clearly the paucity of the services the traditional psychiatrist, who has the same length of professional training period as that of a radiologist.

LIST II

MICHIGAN

RELATIVE

VALUE

STUDY

(1971–1972)

MODIFIERS FOR THE MICHIGAN RADIOLOGIC SOCIETY'S SUGGESTED RELATIVE VALUES FOR CURRENT PROCEDURAL TERMINOLOGY GENERAL INFORMATION AND INSTRUCTIONS

1-General: Subject listings apply only when radiological services are performed by or under the responsible supervision of a physician.

The following definitions apply:

Limited examination: Subject examination usually includes anterior-posterior (AP) and lateral views but less extensive procedures than for a "complete examination."

Complete examination: Subject studies include all necessary roentgenograms for the effective examination of parts involved; additional procedures, exceeding the range of a usual study as fluoroscopy, tomography, cineradiography may be listed separately.

CURRENT PROCEDURAL TERMINOLOGY (C.P.T.)
(Relative Values Recommended by The Michigan Radiologic Society)

		VALUE
70000	Delete number	
70002	Pneumoencephalography, complete procedure	36
70003	Pneumoencephalography, supervision and interpretation only	21
70010	Myelography, posterior fossa, supervision and interpretation only	15
70011	Myelography, posterior fossa, complete procedure	25
Chest		
71000	Chest, "minifilm"	1
71010	Chest, single view	2
71020	Chest, two views	3
71025	Chest, stereo	3
71030	Chest, complete, minimum of four views	4
71034	Chest, complete, minimum of four views, including fluoroscopy (independent chest fluoroscopy, see 76000)	5
74741	Hysterosalpingography, complete procedure	11
74760	Pneumography, pelvis, supervision and interpretation only	8
74761	Pneumography, pelvic, complete procedure	11
74770	Intrauterine contrast injection for fetal study, supervision and interpretation only	8
74771	Intrauterine contrast injection for fetal study, complete procedure	14

List II demonstrates the plethora and sophisticated services of the radiologist and radiotherapist. The list is very much abridged.

LIST III

MICHIGAN INSTITUTE OF PSYCHOSYNTHESIS

RELATIVE VALUE STUDY
March 1972

In compliance with the following excerpt from Federal Regulations-Circulars-Administrative Orders, this Fee Schedule of principal services in available upon request:

Par. 300.19 (d) *PRICE SCHEDULES*: Each non-institutional provider of health services shall maintain . . . a schedule showing its base prices for its principal services. . . . The schedule shall be made available for public inspection, and a copy shall be furnished . . . upon request. Each provider shall post a sign in a prominent place . . . stating the availability and location of the schedule. No price may be increased before the sign is posted and the schedule is made available.

CODE	*TYPE OF SERVICE*	*SESSION*	*PERSONS*	*VALUE*
E	Evaluation, TV-linked	Full	1	6
P	Psychotherapy, TV-linked	Full	1	5
P	Psychotherapy, Regular	Full	1	9
C	Couch Therapy, Regular	Full	1	9
C	Couch Therapy, Primary Telefusion	Full	1	10
C	Couch Therapy, Ordinary Telefusion	Full	1	5
M	Marital or Family Therapy, Regular	Full	2	10
F	Family Therapy, Regular	Full	3	11
G	Group Therapy	1½ hours	1	2
TVB	TV Booster Session	Special	1	2
PT	Psychological Testing, Routine	Full	1	5
EEG	EEG, Routine	Special	1	8
H	Hypnotherapy	Full	1	10
N	Neurological Examination, Regular	Special	1	7
N	Neurological Examination, TV-taped	Special	1	9
	Short Combined Consultation Session			2 extra

EVALUATION

CODE	*TYPE OF SERVICE*	*SESSION*	*PERSONS*	*VALUE*
E-O-1	Evaluation, Ordinary TV-linked	Full	1	6

E-O-2	Evaluation, Ordinary TV-linked	Full	2	7
E-O-3	Evaluation, Ordinary TV-linked	Full	3	8
E-P-1	Evaluation, Primary TV-taped	Full	1	10
E-P-2	Evaluation, Primary TV-taped	Full	2	11
E-P-3	Evaluation, Primary TV-taped	Full	3	12

DIAGNOSTIC SERVICE

CODE	*TYPE OF SERVICE*	*SESSION*	*PERSONS*	*VALUE*
PT	Routine Psychological Testing OIT + MMPI computer	1½ hrs.	1	7
PT-C	Routine Psychological Children under 16 yrs.	Full	1	5
EEG	Routine EEG	Full	1	8
EEG-C	Routine EEG Children under 5 yrs.		1	10
EEG-R	Routine EEG, Interpretation & Reading	Special	1	4
EEG-S	Sleep EEG		1	10
EEG-L	EEG Localization		1	11
TV-PI	TV-linked Personality Inventory	Full	1	5
TV-RP	TV-linked Reactive Psychogram	Full	1	7

PSYCHOTHERAPY

CODE	*TYPE OF SERVICE*	*SESSION*	*PERSONS*	*VALUE*
P-L-1	Psychotherapy, TV-linked	Full	1	5
P-T-1	Psychotherapy, TV-taped	Full	1	9
P-PT-1	Psychotherapy, Primary Telefusion	Full	1	10
P-OT-1	Psychotherapy, Ordinary Telefusion	Full	1	6
P-BT-1	Psychotherapy, Booster Telefusion	Full	1	4
P-L-2	Psychotherapy, TV-linked	Full	2	6
P-L-3	Psychotherapy, TV-linked	Full	3	7

COUCH THERAPY

CODE	*TYPE OF SERVICE*	*SESSION*	*PERSONS*	*VALUE*
C-R-1	Couch Therapy, Regular	Full	1	9
C-PT-1	Couch Therapy, Primary Telefusion	Full	1	10
C-OT-1	Couch Therapy, Ordinary Telefusion	Full	1	5
C-BT-1	Couch Therapy, Booster Telefusion	Full	1	4
C-OT-1+FB	Couch Therapy, Ordinary Telefusion with Feedback	Full	1	7
C-OT-1+FB	Couch Therapy, Booster Telefusion with Feedback	Full	1	6
C-OT-1+P	Couch Therapy, Booster Telefusion Full with programmer as of 8-22-75	Full	1	6
C-OT-1+FB	Couch Therapy, Booster Television Full with Primary Therapist (6-22-76)	Full	1	8

MARITAL THERAPY

CODE	*TYPE OF SERVICE*	*SESSION*	*PERSONS*	*VALUE*
MT-R-2	Marital Therapy, Regular	Full	2	10
MT-PT-2	Marital Therapy, Primary Telefusion	Full	2	11
MT-OT-2	Marital Therapy, Ordinary Telefusion	Full	2	7
MT-BT-2	Marital Therapy, Booster Telefusion	Full	2	5

FAMILY THERAPY

CODE	*TYPE OF SERVICE*	*SESSION*	*PERSONS*	*VALUE*
FT-R-2	Family Therapy, Regular	Full	2	10
FT-R-3	Family Therapy, Regular	Full	3	11
FT-R-4	Family Therapy, Regular	Full	4	12
FT-L-2	Family Therapy, TV-linked	Full	2	6
FT-L-3	Family Therapy, TV-linked	Full	3	7
FT-L-4	Family Therapy, TV-linked	Full	4	8
FT-PT-2	Family Therapy, Primary Telefusion	Full	2	11
FT-PT-3	Family Therapy, Primary Telefusion	Full	3	12
FT-PT-4	Family Therapy, Primary Telefusion	Full	4	13
FT-OT-2	Family Therapy, Ordinary Telefusion	Full	2	7
FT-OT-3	Family Therapy, Ordinary Telefusion	Full	3	8
FT-OT-4	Family Therapy, Ordinary Telefusion	Full	4	9
FT-BT-2	Family Therapy, Booster Telefusion	Full	2	5

FT-BT-3	Family Therapy, Booster Telefusion	Full	3	5
FT-BT-4	Family Therapy, Booster Telefusion	Full	4	7
FT-X-1	Family Therapy, Extra person	Full	1	1

GROUP THERAPY

CODE	*TYPE OF SERVICE*	*SESSION*	*PERSONS*	*VALUE*
G-R-1	Group Therapy, Regular	1½ hrs.	1	2
G-R-F2-3	Group Therapy, Family of 2	1½ hrs.	2	4
G-R-X-1	Group Therapy, Additional member or guest—3 or more	1½ hrs.		1
G-R-Cl-12	Group Therapy, Children under 12	1½ hrs.		2
G-R-Cl3+	Group Therapy, Children 13 or above	1½ hrs.	1	2
G-PT-1	Group Therapy, Primary Telefusion	1½ hrs.	1	2
G-OT-1	Group Therapy, Ordinary Telefusion	1½ hrs.	1	2
G-BT-1	Group Therapy, Booster Telefusion	1½ hrs.	1	1
G-BT-1	Group Therapy Booster on Thursday	1 hour		1
G-LL-1B	Group Therapy, TV-linked-live, booster conclusion	1½ hrs.	1	3
G-LL-2B	Group Therapy, TV-linked-live, booster conclusion	1½ hrs.	2	4
G-LL-1P	Group Therapy, TV-linked-live, personal conclusion	1½ hrs.	1	5
G-LL-2P	Group Therapy, TV-linked-live, personal conclusion	1½ hrs.	2	6
G-A-1	Group Assistant ($5 paid to him)	1½ hrs.	1	No charge

TV BOOSTER THERAPY

CODE	*TYPE OF SERVICE*	*SESSION*	*PERSONS*	*VALUE*
TV-B-1	TV Booster, Regular	Half	1–2	2
TV-B-2	TV Booster Session, Above 2	Half		1
TV-BE-N	TV Booster Evaluation (New Folder + 1 & 2)	Full	1	3
TV-BE-1	TV Booster Evaluation (3, 4, 5 or 6, 7, 8)	Half	1	2
TV-BE-E	TV Booster Evaluation, Double (3, 4, 5 & 6, 7, 8)	Full	1	4
TV-BE-C	TV Booster Complete Evaluation (New folder + 8–8)	2 hrs. Special session	1	8
TV-BL-1-2	TV Booster Long Session	Full	1–2	3
TV-BL-2+	TV Booster Long Session, Extra	Full	1	1
TV-BA-1	TV Booster Assistance (Requiring aid for Booster session)			1

MISCELLANEOUS

CODE	*TYPE OF SERVICE*	*SESSION*	*PERSONS*	*VALUE*
E-N-P	Evaluation, Neurological, TV-Taped		1	9

CODE	*TYPE OF SERVICE*	*SESSION*	*PERSONS*	*VALUE*
LM-0	Brief Medical		1	1
LMT-1	Limited Medical Therapy	Half	1	3
LMT-2	Limited Medical Therapy	Half	2	5
BP-1	Brief Psychotherapy	Special	1	2
BP-2	Brief Psychotherapy	Special	2	4
I-TVT-1	Interview, TV taped, regular	Full	1	9

C-TVT-1	Conference, TV taped, classical	Full	1	9
C-TVC-1	Consultation, TV taped, classical	1 hour	1	20
CV-1	Check-up, TV linked	Full	1	5
CV-2	Check-up, TV linked	Full	2	6
CV-3	Check-up, TV linked	Full	3	7
CV-4	Check-up, TV linked	Full	4	8

List III demonstrates clearly the potential of videological psychiatry and medicine and the range of services, which can catch up with and surpass the level of modern range of service in radiology and radiotherapy with the modern advent of videotechniques in eclectic psychiatry, and family medicine.

By the use of technology, the eclectic psychiatrist can offer a wide range of sorely needed services to the people by becoming a video-psychiatrist.

19. VIDEOTECHNIQUES AND VIDEOSYSTEMS

Videotherapy requires extensive use of TV technology and CCTV system equipment in order to realize a practical clinic to serve the needs of people in search of mental health and family psychiatry and medicine.

Most people are familiar with home television and cable TV, but the practicing psychiatrists and physicians must learn to use the basic elements of CCTV as clinical tools of everyday practice of their clinical science and art of medicine. The basic elements of videosystem are composed of (1) the videocamera; (2) the videocorder (recorder) with videotape; and (3) the videomonitor (TV). This system of three elements forms the one-way input-stayput-output system. The stayput is the videotape, which stores the original information permanently and is available for instant feedback replay. The playback is known as the throughput, that is the picture-on-the-tube.

The basic one-way system of three elements gives us television. And when two such one-systems are linked into a feedback system of two-way information flow like the ordinary telephone conversation or picture-phone, then we have *telefusion*. And videotherapy is based mostly on telefusion. Let us describe briefly the basic elements of the one-way system first.

The videocamera is a TV camera which stores the visual image on a light-sensitive plate, which is then scanned with a series of horizontal lines. The different light intensities on the plate are converted into electrical voltages, which are then transmitted to the next element of the videosystem, that is the videocorder or the video recorder (VTR). The videocorder has a running magnetic-sensitive videotape, which stores the varying voltages of the input image as a train of electrical pulses which order the videotape to respond with fidelity of the original picture in linear form. And when the videotape is played through the videocorder, the magnetic pattern generates a train of electrical pulses identical to the one of input-image, which reappears on the TV screen of the video monitor that is also a light-sensitive plate. However, the picture appears just like on an ordinary TV picture tube. Now, in videotherapy, the therapist is transmitting information (picture of himself and what he says) to the patient, who is seated in another room transmitting information (picture of himself and what he complains of), so that we need two basic videosystems to form a telefusion system to create therapeutic videotapes. What are the fundamental requirements to produce reasonable flow of information in visual and audial forms?

First of all, there must be light in both rooms, but the sound system must be closed in one room (that is, the therapist has to wear an earphone to hear from the videomonitor). Medical TV Consultants, a professional firm in Michigan, advises the use of ordinary neon lights of the tube type with a waffle grill cover to eliminate side-wise lighting from the ceiling. The light from the window has to be draped, otherwise the daylight blinds the videocamera. Secondly, the audio signal is best transmitted in a quiet and well-carpeted room with either a sound-spot microphone in vis-à-vis psychotherapy or couch therapy sessions or to use an omnidirectional microphone in marital or family therapy sessions; and for the group therapy sessions, Medical TV Consultants advise the use of a unidirectional microphone to be passed from speaker to speaker or to use a group assistant who reaches the microphone within six to eight inches from the speaker's mouth.

With the proper lighting and audial environment, we are ready to turn the videocorder on to record picture and sound from the camera/microphone or directly from a TV set of a home TV screen. Now let us refresh our minds about the ordinary TV pictures, which are actually made up of 525 lines of the American standard. A "picture" of 525 lines, if recorded linearly, would require a great deal of videotape, for example, one minute of videocording would require 2,000 feet of tape! However, with modern electronic technology, the lines are recorded diagonally on the videotape in such a way that a great deal of information can be stored by helical scanning. Almost all modern video recorders are based on helical scanning, and this means the videotape must be in excellent alignment in the tape path in order to have good recording on contact with the recording head, which spins at about 1,800 rpm inside the recording drum. The video recording head makes contact with the videotape, so it is subject to wear and tear. And the recording is easily damaged, when tape threading is done with the head spinning on the reel-to-reel machines. Medical TV Consultants therefore advises the use of cassette type.

The NV-cassette model is a one-half-inch video cassette recorder of EIAJ color standard with black/white capability. The one-half inch recorders are priced at about $1,000 and the one-half inch thirty-minute videotapes cost at about twenty dollars per tape. Medical TV Consultants advise the use of one-half-inch machines and tapes for practical clinical economy and efficiency. The choice of the initial system is important in order that the system can grow from television to telefusion and to multiple-input-out-videosystem, with compatibility to link with other videotherapists as well as with the ready availability of service.

In order to secure good and reliable video equipment and videosystem, it is important to obtain information from a respected firm in electronics with engineering capabilities in all three component areas: high-performance CCTV cameras, precision videotape recorders, and reliable video monitors. If the reader needs information in medical videology, the following illustrations of the components with descriptions and prices may be consulted as supplied by Medical TV Consultants, Michigan.

CCTV CAMERAS MODEL WV-361P

⅔″ VIDICON CAMERA WITH 6″ BUILT-IN MONITOR

WV-361P embodies the latest electronic developments. Among the WV-361P's quality features is a four-position monitor selection switch that allows the camera viewfinder to function as a VTR record monitor, a VTR playback monitor and a line view monitor. A camera operator in a multi-camera/VTR production system can now view the program output in addition to the output of his camera. The large 6″ diagonal viewfinder displays brilliantly clear, sharp pictures. The large, top-mounted tally light is visible from any position to both operator and performer. A number of sync modes makes the WV-361P suitable for a variety of uses . . . internal: random or plug-in 2:1 interlace board (optional); external; EIA 2:1 interlace or Panasonic pulse generator.

All necessary camera controls are within easy reach. Internal/external sync switch, optical focus, viewfinder selection switch, on/off phone volume control, pilot lamp, headphone and intercom jack. Viewfinder controls . . . brightness, contrast, horizontal, and vertical hold are all easy to adjust at the rear of the camera. Beam, focus, and auto/manual target level controls are on the side of the camera. The camera will operate with as little illumination as 2 ft.-candle (20 lux).

The WV-361P's housing is designed to accommodate an optional, pushrod type rear control zoom lens.

FEATURES

1. ⅔″ separate mesh vidicon.
2. Quadruple function viewfinder (camera viewfinder, VTR record monitor, VTR playback monitor, line view monitor).
3. Resolution 550 lines at center.
4. Decreased lag, lower noise, and increased sensitivity.
5. Automatic light compensation circuit accommodates variations of 5000 to 1.
6. Automatic and manual target control.
7. No more blur and better picture quality with White Clipper Circuit.
8. 6″ viewfinder.
9. Internal/external sync switch.
10. Plus-in 2:1 interlace sync board.

11. Built-in intercom and headphone connections.
12. Built-in tally light.
13. Single multiple-pin connector (sync input/video output) for use with Panasonic Special Effects Generator.
14. 3 prong, grounded power cable.
15. High reliability and easier maintenance with FETs, ICs, and silicon transistors.

SPECIFICATIONS

Power Source: AC 117V, 60 Hz
Power Consumption: Less than 31 watts
Vidicon Tube: ⅔″ Separate mesh vidicon 20 PE 13A/8844
Viewfinder: 6″ 70° deflection
Scanning Standard: 525 lines/60 fields/30 frames
Sync: Int Random horizontal, line-locked vertical or plug-in 2:1 interlacing printed circuit board (optional) Ext. 2:1 Interlacing EIA (VD, HD) or Panasonic pulse generator
Video Frequency Response: More than 7 MHz
Horizontal Resolution: 550 lines (center)
Video Output: UHF out: 1.0Vp-p Composite/75 10P out: 1.0Vp-p Composite/75 or 0.7Vp-p Non-Composite/75 switchable
Signal-to-Noise Ratio: More than 40 dB
Automatic Light Compensation: 5000 to 1
Required Illumination: Over 2 ft. candle (20 Lux) (F1.6 lens)
Lens Attached: F1.6 16 mm C-mount (no iris)
Operating Ambient Temperature: 14° F-122° F
Operating Humidity: Less than 90%
Solid State Devices: Transistors: 32 pcs. Silicon Control Switch: 2 pcs. Diodes: 25 pcs. Integrated Circuits: 5 pcs.
Dimensions: 7″ (W) × 7½″ (H) × 14½″ (D)
Weight: Approx. 14.3 lbs.
Standard Accessories: Coaxial Cable 1 pc. F1.6 16 mm C-mount Lens (no iris) 1 pc.

Optional Accessories: Camera Cable 10G-25, 10G-50, 10G-100 Extension Camera Cable 10H-25, 10H-50, 10H-100 WU-V90 Fujinon Zoom Lens Rear Control type: H5-14P (F2.0 14-70 mm C-mount) H10-11P (F2.5 11-110 mm C-mount) Manual Control type: H5-14C (F2.0 14-70 mm C-mount)

VIDEO CARTRIDGE TAPE RECORDERS

MODEL NV-5120

½″ VIDEO CARTRIDGE RECORDERS EIAJ ½″ COLOR AND EIAJ # BLACK/WHITE STANDARD

Tape recorded on any industry standard ½″ reel-to-reel VTR (color or black and white) may be played on NV-5120. Conversely, tape recorded on this cartridge machine may be played or edited on EIAJ reel-to-reel VTRs. No electronic equipment is needed for either conversion.

The unit operates automatically when the cartridge is pushed in. Stop, play, rewind, and fast forward pushbuttons may be used at any time. Functions are performed instantly when pushbuttons are activated. When the tape ends, it is automatically rewound and the cartridge pops out for each removal. A tape indicator enables the user to fast forward to an arbitrary program, selectable by a simple dial setting. Another convenience is the built-in RF adaptor that permits direct playback through conventional TVs.

To assure a color picture free from annoying fluctuation, an Automatic Phase Control circuit maintains the stability of the color signal. A drop-out/noise compensator virtually eliminates loss of video signal by replacing lost video information with matched signals, and reduces the annoying random electrical interference that could otherwise produce a grainy effect on the screen.

FEATURES

1. EIAJ color and #1 b/w standard VTR.
2. For record/playback on ½″ tape cartridges.
3. Built-in RF adaptor permits playback through conventional TVs.
4. Permits simple mechanical conversion to cartridges of any tapes made on EIAJ open reel VTRs.
5. Permits simple mechanical conversion of cartridge to open reel for use on any EIAJ VTR.
6. Dropout/noise compensator virtually eliminates dropout.
7. Automatic Phase Control circuit assures high stability of color image.
8. One step operation—simply insert the cartridge.
9. When program ends, tape rewinds automatically and cartridge ejects.
10. Simple dial setting permits fast forward to preselected program.

11. Still framing.
12. Fast forward and rewind.
13. Audio dubbing.
14. HPF* heads afford extra long head life and assure high picture quality.
15. Precision inscribed tape path assures 100% interchangeability with all EIAJ standard VTRs.
16. Skew and tracking controls.
17. Automatic end of tape shutoff.

SPECIFICATIONS

Power Source: AC 120V, 60 Hz
Power Consumption: Approx. 110 Watts
Video Recording System: 2 rotary heads, USA standard (525 lines, 60 fields)
Video Modulation System: Frequency modulation
Tape Speed: 7½ I.p.s.
Tape Width: ½ inch
Heads: Video: 2 rotary heads Audio/Control: 1 stationary Erase: 1 for both video and audio 1 for audio dubbing
Recording Time: 30 min. with NV-P530 (1,200 ft.)
Horizontal Resolution: (on Monoscope Test Pattern) Color more than 240 lines
Active Elements: Fully transistorized
Frequency Response: Audio: 50 12,000 Hz
Signal-to-Noise Ratio: Video: more than 40 dB Audio: more than 42 dB
Input Level: Video: 1.0 Vp-p Audio: MIC more than—72 dB AUX more than—30 dB
Input Impedance: Video: 75 unbalanced Audio: MIC 600 10 k unbalanced AUX 220 k unbalanced
Output Level: Video: 1.0 Vp-p (0 dB) Audio: 1.0 v rms
Output Impedance: Video: 75 unbalanced Audio: more than 2 k
Dimensions: 221/16″ (W) × 15¾″ (D) × 8½″ (H)
Weight: Approx. 52.8 lbs.
Standard Accessories: EC cord VTR-monitor connection cable

VIDEO CARTRIDGE TAPE (OPTIONAL) SPECIFICATIONS

Format: EIAJ standard one reel cartridge tape
Tape Width: ½″
Tape Length: NV-P510: 400 feet (10 min.) NV-P520: 800 feet (20 min.) NV-P530: 1,200 feet (30 min.)
Cartridge Dimensions: 5″ (W) × 5½″ (D) × 1⅛″ (H)
Cartridge Weight: NV-P510: Approx. 10.5 oz. NV-P520: Approx. 12.3 oz. NV-P530: Approx. 14.1 oz.

CCTV MONITORS

MODEL TR-910M

BLACK AND WHITE MONITOR

This presents a compact, lightweight Solid State CCTV monitor featuring thirty-eight square inches of viewing area and reliable, all transistor circuitry. TR-910M delivers over 450 lines of horizontal resolution for the finest closed circuit television reception. Its 90° aluminized picture tube insures sharp, clear pictures. TR-910M's controls are front mounted for easy operation. TR-910M operates on AC or can be battery operated from 12V outlet.

SPECIFICATIONS

Power Source: AC 120 V, 60 Hz, 26 W (DC 12V, 12W)
Video Input: 1.4 Vp-p 75 High Z UHF Type Bridging
Resolution: Video: More than 450 lines
Picture Tube: 90° deflection aluminized 38 sq. in. viewing area
Weight: Approx. 15 lbs.
Cabinet Finish: Black with silver trim diecast handle
Dimensions: 9⅝″ (W) × 95/6″ (H) × 9″ (D)

VTR MICROPHONES

WM-1241P

The WM-1241P is an omnidirectional dual-impedance dynamic microphone. The WM-1241P also features an on/off switch and a windscreen. The windscreen makes the WM-1241P suitable for outdoor use while the on/off switch is helpful whenever intermittent amplification or recording is desired. Includes a screw adaptor, a stand adaptor, and a desk stand.

WM-1340P

The WM-1340P is a unidirectional dynamic microphone suitable for musical as well as voice applications. Complete with a stand adaptor and 16′5″ of cable, it features an on/off switch and a dual-impedance phone plug (250 or 20k). The WM-1340P's built-in windscreen also helps eliminate annoying "pop" caused by pronouncing the letters "b" and "p."

Because of its cardioid directional response, it can be used in close proximity to speakers without feedback problems and is less susceptible to the echo or "boominess" associated with some speaking locations.

WM-1244P

The WM-1244P is an omnidirectional dual-impedance dynamic microphone. The WM-1244P also features an on/off switch and a flexible gooseneck shaft. Useful for general applications, including those that require intermittent recording or amplification, the WM-1244P gooseneck easily adjusts to any desired height and direction (stand WN-1112P shown optional).

WM-1600P

Model WM-1600P is a self-biased omnidirectional condenser microphone employing an electret foil diaphragm. It has a frequency response of professional quality and is a suitable transducer for high fidelity sound applications. Field effect transistors in the impedance conversion circuit give the WM-1600P a battery life of about *10,000* hours. It is supplied with a removable polyurethane windscreen that protects against wind noise and "pop." Although it has a low output impedance (250), its high sensitivity makes the WM-1600P suitable for high input impedance (20k) amplifiers. Comfortable when hand-held, the WM-1600P also comes with a screw adaptor, a stand adaptor, a desk top stand, and an "AA" size alkaline battery.

WM-1650P

Model WM-1650P is a unidirectional electret condenser microphone with transducer properties similar to those of the WM-1600P except for directional characteristics. A cardioid polar response enables the WM-1650P to be used near loudspeakers with fewer feedback problems and makes it less susceptible to the "boominess" associated with some halls. It comes with a screw adaptor, a stand adaptor, a desk top stand, a removable polyurethane windscreen, and an "AA" size alkaline battery.

Model Number	Description	Approximate Shipping Wt. (lbs.)	Suggested Retail Price
	VIDEO TAPE PLAYERS—CARTRIDGE and CASSETTE		
NV-5110 NV-2110	½″ Video Cartridge Player; color and black & white; compatible with EIAJ #1 black & white and EIAJ ½″ color standards. Automatic tape threading, end-of-tape rewind and cartridge ejection, patented HPF magnetic heads, still picture, pre-set auto repeat, Automatic Phase Control (APC).	50.0	1050.00
	¾″ Video Cassette Player; color and black & white, 2-track audio, automatic stop at end of tape, elevator tape loading system.	66.0	1150.00
	VIDEO TAPE RECORDER/PLAYERS—CARTRIDGE AND CASSETTE		
NV-5120	½″ Video Cartridge Recorder/Player (VCR); color and black & white, compatible with EIAJ #1 black & white and EIAJ ½″ color standards. Automatic tape threading, end-of-tape rewind and cartridge ejection, patented HPF magnetic heads, still picture, pre-set auto repeat, Automatic Phase Control (APC).	60.0	1350.00
NV-5125	½″ Video Cartridge Recorder/Player (VCR) with built-in VHF/UHF tuners for off-air recording, color, and black & white, compatible with EIAJ ½″ color standards. Automatic tape threading, end-of-tape rewind and cartridge ejection, patented HPF magnetic heads, still picture, pre-set auto repeat, Automatic Phase Control (APC).	60.0	1450.00
NV-2120	¾″ Video Cassette Recorder/Player; color and black & white, 2-track audio, audio dubbing, AGC on audio and video, automatic stop at end of tape, elevator tape loading system.	[illegible]5.0	1450.00

Model Number	Description	Approximate Shipping Wt. (lbs.)	Suggested Retail Price
	VIDEO TAPE PLAYERS—OPEN-REEL		
NV-3010	EIAJ #1 black and white standard video player in portable case; single lever operation.	38.0	595.00
NV-3110	EIAJ ½″ color standard and EIAJ #1 ½″ black and white standard video player in portable case; single lever operation, dropout noise compensator.	45.0	895.00
	VIDEO TAPE RECORDER/PLAYERS—OPEN-REEL		
NV-3020SD	EIAJ #1 black and white standard video recorder in portable case, slow motion on playback, electronic editing permits positive editing of recorded tape electronically, AGC on audio and video, automatic end of tape shutoff, audio dubbing.		

CCTV CAMERA
MODEL WV-361P

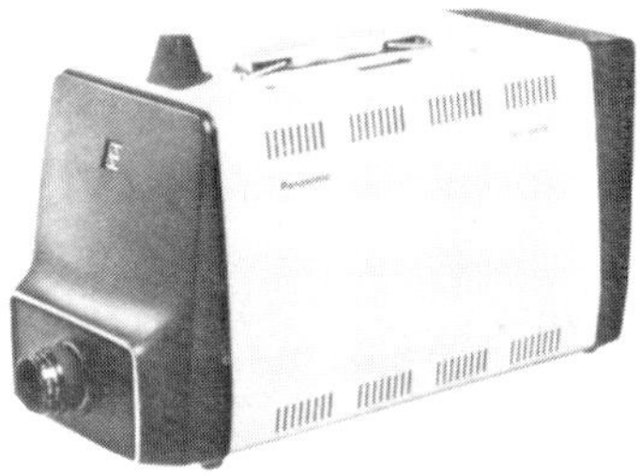

VIDEO CARTRIDGE
TAPE RECORDERS
MODEL NV-5120
MODEL NV-5120E

VIDEO TAPE RECORDER
MODEL NV-3020SD

VIDEO CARTRIDGE TAPE RECORDER
MODEL NV-5125

VIDEO CASSETTE TAPE RECORDER
MODEL NV-2125

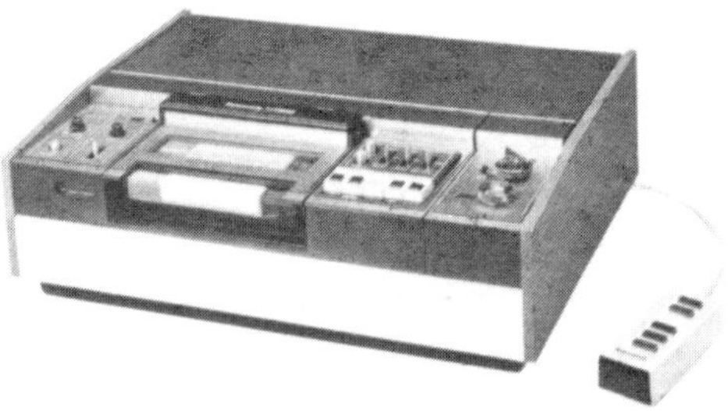

Video Cassette Recorder
shown with optional NV-A150

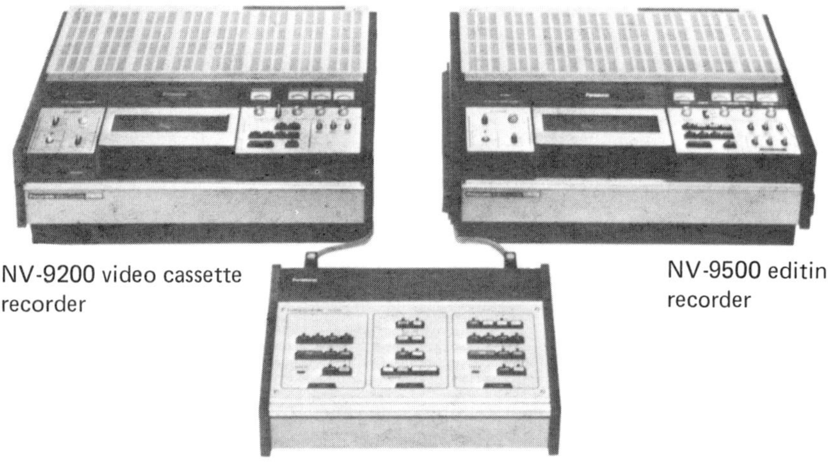

NV-9200 video cassette recorder

NV-9500 editing recorder

NV-A950 editing controller

VIDEO CASSETTE EDITING SYSTEM

MICROPHONES

WM-1240P

The *WM-1240P* is an economical dynamic microphone. This omnidirectional, dual impedance microphone is designed for versatility in non-musical applications. Designed for comfortable hand held use, *WM-1240P* can also be used on a stand; a screw adaptor, stand adaptor and desk stand are included.

WM-1241P

The *WM-1241P* is an omnidirectional dual-impedance dynamic microphone.

WM-1244P

The *WM-1244P* is an omnidirectional dual-impedance dynamic mincrophone.

WM-1340P

The *WM-1340P* is a unidirectional dynamic microphone suitable for musical as well as voice applications.

20. VIDEOLOGY AND THE FUTURE PRACTICE OF MEDICINE

Videology is the science and art of medical television. Videology is analogous to radiology, in that here we employ the use of camera and tape for the examination and diagnosis and therapy in the treatment and prevention of disease and safeguarding of health, as we have in radiology in the employment of X-ray machine and film.

Historical Comments. Videology came out of psychosynthesis. In the classical psychosynthesis of Assagioli, many minds have been striving for a fusion of technique in psychotherapy and psychiatry. This search for a synthesis has been ridiculed and rejected since the historical quarrel between Freud and Adler. Adler, in the back of his mind, attempted to bring forth his ideas of class problems, society, into the individual psychoanalysis. Freud severely criticized Adler for using psychoanalytic concepts hidden in new Adlerian phraseology. The early psychoschisis between Freud and Adler and their conflicts was rooted in the struggle: individual versus community approach. However, modern television technology may bring a new synthesis to this conflict.

Medical Television Techniques. Medical television techniques as developed by psychosynthesis will not only transform the practice of psychiatry but also of medicine itself. Medical television is still at its early stages of development. In academic circles and in the universities, television has been used in the last two decades for the purpose of medical education. The techniques used in medical schools have, by and large, followed the common-sense approach of TV studios, without definitive innovating in techniques relevant to medical practice. It has become apparent to me that videology can transform the passive viewing instrument of TV camera and tape into active elements of medical examination and therapy. A basic medical TV system can help the individual physician to secure medical history to educate patients for preventive measures, for prenatal care, preoperative and postoperative treatment programs to be carried out by the patient, to assist relatives in diet programs or insulin injection techniques, to prepare good medical records, and for the use of medico-legal deposition and forensic evidence. The uses and techniques are simply open-ended and limitless. The art and the science of videology can be made so flexible and broad by the expertise of the individual physician, general or specialist, and to the needs of his patients, for curative or preventive purposes.

In fact, videology has developed so extensively in the practice of psychosynthesis that individual couch therapy can be applied on a mass scale. Here we begin to appreciate the application of cybernetic medical-TV techniques. We can produce almost standardized psychoanalysis and psychosynthesis, ad lib. The mass potential of offering individual psychoanalysis and psychosynthesis to the general public in the world community has become a distinct possibility with TV-cybernetic technique in psychosynthesis: telefusion.

Telefusion. Telefusion is the bringing together of information from two sources, forming a feedback system. Television is publicly experienced as one-way communication, namely from the television studio to the television receivers at home. Telefusion is two-way television communication. With two-way TV communication, we begin to appreciate the possibilities of substituting the traditional eyeball-to-eyeball approach in the practice of medicine or person-to-person approach in the practice of psychoanalysis, in which the patient becomes wholly dependent on the presence of the therapist, serving often as a high-priced monitor, wherein therapeutic process and efficacy are so difficult to measure. In the practice of telefusion couch therapy, we have new freedoms of choice. Couch therapy in different telefusion modes can be offered:

(a) Active-active mode.
(b) Active-passive mode.
(c) Active-passive-active mode.
(d) Active-passive-passive mode.

In the *active-active mode* (a), the patient and the therapist are both on television monitors, and their communication is reciprocally communicated with instantaneous feedback as in classical couch therapy.

In the *active-passive mode* (b), the patient stays active on television monitor and interacts with the passive therapist on television, but the therapist's passivity is standardized by having him pre-taped or pre-programmed. There we have created time and standardized therapy in telefusion couch therapy. This naturally gives rise to the third mode, active-passive-active.

In the *active-passive-active mode* (c), the patient can be active, interacting with the standardized passive therapist on television, pre-programmed. At the end of the session or on another day, the delayed feedback or the need for interaction is actively carried out by the therapist's personal participation at a more convenient time or situation as therapeutically and clinically defined.

In the *active-passive-passive mode* (d), the patient passively watches his own feedback on couch. The patient can sit passively where the psychoanalyst traditionally sits since the days of Freud.

These modes show the potential of cascading, or interlacing, in many TV-linked therapeutic techniques. In fact, the individual therapist can design however complex a situation he wishes to. For example, we could do cross-lateral telefusion couch therapy, that is to say, a woman can produce free association on couch, which then is fed forward to her husband, and the husband's free association on couch can be fed forward to the wife, and then their productions can be integrated and synthesized with the therapist in a family session. In this cross-lateral telefusion couch therapy, many variables arise and clinical judgment must be exercised cautiously. I am here illustrating a complex mode to demonstrate the magnitude of therapeutic power in telefusion, and not necessarily recommending as such.

At this point, a question may be raised on what theoretical model can one carry out such telefusion techniques? I would say simply on the theory and principles of psychosynthesis, in which we view free association of classical psychoanalysis as a random output theorem, namely, the patient is producing material "at random." His inhibition is reduced and he is encouraged to bring out hidden grievances, blocked communication of thought, emotions and ideations, communication with the purpose of feedback and correction in an individual's life. In effect, free association is a form of free communication, except viewed as "random output." Psychosynthesis has given rise to the concept of free communication, namely, the therapist can also produce "random output," viewed from the angle of the patient's "random input," into the patient's mind to provide thoughts, ideas and corrective imageries so that the patient obtains the proper information needed for self-reorganization, reconstruction, and higher personality synthesis.

Telefusion couch therapy will become a standard technique in which two-way TV-cybernetic system employing free association of classical psychoanalysis with "random output" is coupled with free communication of cybernetic psychosynthesis with "random input." By using the active-passive mode, history may be obtained in psychosomatic disorders from any patient by family physicians, internists, pediatricians, or whichever psychodynamics is important for the treatment and prevention of physical and mental disorders. The free communication techniques can of course be used in preventive measures whether they be regarding inoculation, prenatal care, adolescent adjustment, sex education, and so forth, in which the family physician uses his television personality to explain to the families and family members in his care.

In short, videology will become such an integrated part of any future-oriented practice of medicine that the personal touch is not only enhanced but amplified many-fold. To grasp the future practice of medicine in terms of videology is to conceive man as the highest synthesis of information. The best personality is the most integrated synthesis of the widest information about the individual, his family, his community, and the world community, not only from historical perspective but also with futuristic goals.

Videology will become an essential part of the practice of medicine when television is not viewed as a passive viewing box but actively sought as a cybernetic system, an extension of the physician's brain, voice box, and his professional personality, in the service of all those who come to him where the economic values of class struggle will be gradually removed and dismantled. In videology there exists the creative acorns of the future practice of medicine.

I would hope that videology will be taught as a separate discipline in medical schools and departments of videology established in hospitals, clinics, and wherever medical centers are preparing for the future and regard medical care as a universal right of all people.

A PSYCHOSYNTHETIC GLOSSARY

A

AMBIGRAM: A written symbol which can be perceived as two different words. For example:

The coexistence of two words in the same set of written stimuli. The  barks and the  cries.

AMBIGRAPH: A picture which can be perceived as two objects, e.g., a bird and a rabbit.

The coexistence of two perceptual pictures in one set of stimuli.

AMBISTORY: A story which can be seen as two different stories. For example, father and son were in a car wreck. The father was killed outright. The son was taken to the emergency room of the hospital. The surgeon took one look and exclaimed, "Gosh, this is my son. I cannot operate on him!" What happened? For answer, see footnote below, written backwards.*

AMBIVALENCE: The coexistence of two opposite feelings for the same object or person. For example: a boy may love and hate his father out of a complex of ambi-experiences.

AMNESIA: Loss of memory. This is usually meant in the sense of pathological loss. However, infantile amnesia is physiological lack of recall. In psychosynthesis, amnesia is electrically induced for the purpose of reprogramming.

ANALYSAND: One who is undergoing psychoanalysis.

*Footnote: Eht noegrus si eht s'yob rehtom.

AND-RULE: The and-rule states that A-and-B is true only if both A and B are separately true. For example: if a child can go to a movie only when both father and mother separately vote yes, then the family follows the and-rule.

B

BIT: Binary digit is a bit, the smallest measure of information. A bit is the quantity of information which can be measured as a binary choice of yes-or-no, on-or-off, etc. For example: the information contained in the toss of one coin to reach a decision, based on head-or-tail, constitutes one bit.

BIT: Contraction of binary digit or bit, the smallest unit of information. In psychosynthesis, a bit represents a pulse fired by a neuron in a matrix of pulses fired by the neurons in relation with input or output information flow.

BINARY CODE: The number system which uses 2 as its base. The code is either zero (0) or one (1), representing "on" or "off," "yes" or "no," "firing" or "not-firing" as in neuron or brain cell activity. In psychosynthesis, consciousness is represented by the matrix of "on" or "off" of neuron activity. In electrolytic therapy, ELT, it is postulated that all neurons are set off by electricity to firing, and as a result consciousness disappears and convulsion appears as a by-product. And then the neurons go into a resting phase, which emerges as a more programmable state.

BOOLEAN ALGEBRA: An algebra invented by George Boole (1815–1864), who described logic in terms of AND, OR, NOT, etc., operations based on binary system.

C

CATALYZER: A person who helps to reprogram a patient or synthesant in the postlytic session. For example: If a wife is undergoing psychosynthesis, the husband will serve as the postlytic catalyzer.

CHANNEL: The pathway along which information flows. In psychosynthesis, channel may be physical, chemical, electromagnetic, or combinations thereof. In this unified theory, information may be modified physically, chemically, electrically, or combinations thereof without contradictions.

CHEMOLYTIC: The process of loosening or erasing memory and consciousness by chemical means. For example: the use of indoklon in indoklonlytic therapy or the use of LDS in model psychosis or drug abuse.

CIRCULATING MEMORY: Moving and reverberating information on the background neuronal discharges, providing more immediate recall, retrieval, and reaction under proper input information. When circulating memory is stimulated to discharge thinking, feeling, or overt behavior, scintillating memory emerges. And this neuronal scintillation is consciousness.

CLEAR: To reset the system to its initial state. In electrolytic therapy, to clear is to loosen the scintillating and circulating memory to zero.

CODE: A system of symbols representing information about objects and interaction between objects in space and time. For example: the binary system of neuron firing represents a code of reality of sensation, perception, conception, ideology, consciousness, and behavior.

COFFEE SHOP: The Coffee Shop of psychosynthesis is more than an ordinary coffee shop. It is a coffee shop which serves as a therapeutic stage for personality development and community mental health. In the Coffee Shop, the psychodrama of Moreno has become the drama of everyday life, in which the patients and their family members have actually put their time, money and efforts together to build a coffee shop corporation (Families for Psychosynthesis, Inc.), which sells stocks and provides therapeutic, recreational, occupational, and financial opportunities for all family members. The family members have forgotten the "sick patient" role assigned to them by social convention and operate the coffee shop as a business in public. The family members *act* as president, treasurer, secretary, chefs, waitresses, managers, accountant, or lawyer, and in a variety of roles in individual social relationships.

In brief, the Coffee Shop of psychosynthesis has the same relative therapeutic significance as the couch to psychoanalysis. And the Coffee Shop represents a fusion of techniques.

CONSCIOUSNESS: The scintillating neuronal integration at a given moment of a time-series of input information and output motor impulses with the associating memories of the past for the preparation of continued reception of further sensory input and motor output as required by the brain to maintain its psychophysiological stability for survival in a given environment.

CONTRADICTIONS-IN-IDENTITY, Principle of: One of the three basic principles of psychosynthesis states that

$$\text{If PRI } (E_i, E_s) \geqslant T, \text{ then } E_i = E_s$$

and

$$\text{If PRI } (E_i, E_s) < T, \text{ then } E_i \neq E_s$$

which means that pattern recognition of any object depends on an arbitrary threshold. An object can be viewed as either A or Not-A, depending on the assignment of the threshold function. This can be graphically demonstrated by the Jastrow-Boring Phenomenon (See also Ambigraph).

CONTROLLING PERSONALITY: See Dominant Personality.

COSMAN: Not an individual person, in the sense of Superman of Nietzsche, but the fusion of all mankind with a cosmic identity. Cosman is the evolving synthesis of three billion or more human beings, supported by other forms of life on this planet, linked by mass communica-

tion, mass transportation, and mass participation as a unified ecosystem in the universe. Cosman is based on the ideology of psychosynthesis. Every person then becomes an important element of the living earth on its exciting journey, exploring the galaxies.
CYBERNETICS: The study of feedback and control of information in the machine, animal, man and his environment.

D

DOMINANT PERSONALITY: The controlling time-series of memory, habits, and behavior neural firing pattern of an individual. Analogously, the controlling personality or dominant personality is the ruling party of a given brain at any time interval.
DYNAMIC THRESHOLDS: Principle of: One of the three basic principles of psychosynthesis states that

$$\begin{aligned} &A = A \text{ if } T \geqslant D \\ &A \neq A \text{ if } T < D \\ &D = D \text{ if } D \geqslant K \\ &D \neq D \text{ if } D < K, \end{aligned}$$

which means the object of identity is being transformed continuously into another, even to its opposite, since the threshold function is never stable. We define: A = any object in the universe. T = threshold. D = threshold constant of A. K = threshold constant of D.

E

EGO: The conscious self. In Freudian psychology, ego deals with reality, by coordinating the instinctual forces of the Id and the restrictive superego force of society and conscience. In psychosynthesis, Ego or I is the executive personality at any given moment.

Ego is a frame of the time-series of self-creation, fusing social and environmental input information as well as memory and genetic information of the individual.
ELT: Abbreviation for electrolytic therapy. ELT is the technique of inducing loosening and erasure of memory and consciousness for the purpose of reprogramming personality, by the application of TV-linked electrotherapy with family psychotherapy.
ENERGY: That which performs work. In psychosynthesis, that which changes patterns. Personality is a time-series of energy patterns.
ENERGY OF TRANSFORMATION: The amount of work required to change one pattern into another, or in psychosynthesis, the amount of energy required to transform one personality into another. The form of energy may be verbal, chemical, mechanical, or electrical.

ERASE: To replace the binary digits of information of the system, i.e., brain, to binary zeros as in electrolytic therapy. To erase is to clear in electrolytic therapy.
ERASING: The process of erasure. The subtraction of information from the memory.
ERASURE: The process of erasing information or the procedure to subtract information from a system. All classical procedures and theories of psychoanalysis and psychosynthesis depend on the addition of information into the brain, whereas cybernetic psychosynthesis focuses on electrolytic or chemolytic erasure of information from consciousness and memories as a central practical and theoretical consideration.
ERROR: The difference between goal and actual achievement. It is the discrepancy between identity and most-similarity.

F

FEEDBACK: The return of a small amount of positive or negative information from output to the input of a system. Feedback is the key process of cybernetics for purposive behavior or goal-orientation.
FREE ASSOCIATION: The technique of random output on the couch by the analysand in psychoanalysis of Freud.
FREE COMMUNICATION:The technique of random input to the patient by the therapist via telefusion of psychosynthesis. Free communication completes the loop of free association.

G

GOAL-DIRECTED THERAPY: A therapeutic program, which is based on pre-defined and well-described goals to be achieved. For example: Psychosynthesis is goal-directed therapy, based on the electrolysis of undesirable personality and resynthesis of a desirable personality with a program linked by television, the eclectic elements of modern psychiatry, such as classical analysis, synthesis, chemotherapy, electrolytic therapy (ELT), behavioral therapy, family therapy, group therapy, and community mental health.

I

INFORMATION: The energy pattern or coding representing an aspect of reality.
IMPEDANCE: The collective resistance of others around the patient in preventing the patient from attaining a healthier or higher personality.

J

JASTROW-BORING PHENOMENON: The phenomenon is a psychological demonstration of the truth of the second principle of psychosynthesis, namely, the principle of contradictions-in-identity. That is to say, an ambigraph, such as designed by Jastrow (duck-or-rabbit) or by Boring (beautiful girl-or-ugly witch) can be seen as two different objects, depending on the threshold of perception. And this phenomenon is the basic element of the two-personality theory. (See also Ambigraph, Two-Personality Theory.)

K

KINETIC MEMORY: Moving information. Kinetic memory is either circulating or scintillating. Circulating memory is preconscious and is co-determined both by permanent memory (genetic information) and semi-permanent memory (culture). Scintillating memory is consciousness.

L

LAW OF UNIVERSAL TRUTH: See Universal Truth, Law of The Law of Universal Truth is based on the principle of most-similarity. (Ref.: See H. C. Tien, "The Theory of Pattern Recognition by Identity," *World Journal of Psychosynthesis* 1, no. 3, (November 1969).

LOOSENING: The process to induce weakening of consciousness, associations, and memories by electrical or chemical means. For example: electrolysis is a type of loosening.

LYTIC: Formed from lysis, to loosen. Lytic therapy refers to the loosening of memory and consciousness for personality transformation as in electrolytic therapy (ELT) and chemolytic therapy (CLT). The lytic step of ELT is the actual introduction of current to loosen memory and consciousness.

LYZER: The electrolytic machine used to introduce electrolysis. At the present (1971), the lyzer of ELT is the same equipment as used in ECT. For example: the Reiter SOS Machine.

M

MASTER PROGRAM: The earliest program of a system which directs the subsequent input and output of a system.

MATRIX: A set of elements connected to form a system to process information, i.e., brain cells connected by dendrites and axons.

MEMORY: Stored information, stored energy patterns, representing information about the environment. Memory may be classified as static or kinetic. Static memory is permanent or semi-permanent and kinetic memory is circulating or scintillating.

MIND FUSION: Literal translation of psychosynthesis. The phenomenon of psychosynthesis occurring or existing in one individual's mind, while psychosynthesis connotes the union of all minds. See Psychosynthesis.

MOST-SIMILARITY, Principle of: One of the three basic principles of psychosynthesis states that

$$\text{if PRI } (E_i, E_s) \rightarrow 1, \text{ then } E_i \rightarrow E_s,$$

which means that no pattern can be identified with absolute certainty in the Universe. (PRI = pattern recognition by identity). In fact, there are no identities in the universe. A is never A, $A \neq A$. In. all mathematical reasoning, $A = A$ really means $A \rightarrow A$.

In other words, any object in the Universe may be identified by remembering some energy patterns, which are given off as reflected light, radiated heat, mechanical vibrations, etc., but it can be identified only as most-similar to the stored perception, but never completely.

N

NAME: The symbol of a personality. For example: John is a generous person. John is the code of a time-series in a given person.

NEONAME: Same as transname or transnom. Transnom is more acceptable because it is more euphonic.

O

OR RULE: The logic of A-or-B is true when A or B or both are true. For example, if a boy may go to the movie is either father votes "yes" or mother votes "yes" or both vote "yes," then or-rule is followed.

P

PATIENT-THERAPIST ALLIANCE: The concept of transforming every patient into a therapist, $Pt \rightleftharpoons Th$

PATTERN RECOGNITION: The process of reducing input, or sensory data of a system, to a specific output code, a name. Pattern recognition is a process that gives a complex pattern a simple name. In short, pattern recognition is name-calling. For example: the complex personality of a girl may be simply called Anna. Pattern recognition may be said to be the opposite process of psychosynthesis, which attempts to form a personality pattern from a name.

PATTERN RECOGNITION BY IDENTITY THEORY: Or the PRI

Theory, which states for any system, computer or biological, to recognize a pattern (e.g., A) it must remember the pattern A.

PERF MODEL: The PERF Model (acronym for Program-Edit-Reprogram-Feedback) is a practical derivation of the TOTE Model. The PERF Model can be applied to any psychotherapeutic system, whether it be psychological, family, psychoanalytical, psychopharmacological and/or electrolytical. However, the model originally evolved from electrolytic therapy (ELT) for the purpose of analysis of a personality (Program), followed by electrolytic erasure or modification (Erase) of memory, so that a new, higher personality may be developed (Reprogram) with a therapeutic goal which can be used to compare with the actual therapeutic attainment by cybernetic techniques (Feedback). In the reprogramming, sometimes we use a new name for the new personality identity. This is transnomation. (See also TOTE Model; Transnomation.) (See photographs which illustrate the PERF Model.)

PERSON: The physical set of an individual. A person may have more than one personality. In psychosynthesis, it is theorized that every person has infinite personalities, but by arbitrary threshold and custom, two personalities may be perceived in every person. The dominating personality is in control of the person. The secondary personality may compete for dominance.

Analogously, a person is like a nation with two political parties.

PERSONALITY: Persona = mask. A personality is defined as a time-series of electrochemical scintillations, or neuronal discharges. It is a pattern of input-processing-output relationships of the individual to himself and his environment. In psychosynthesis, the pattern is to be transformed.

PERSONALITY TRANSFORMATION: The process of changing a personality by verbal, chemical or electrical means. Psychosynthesis is the unified approach of personality transformation by using all avenues.

PLEDGE OF PSYCHOSYNTHESIS: The master program is a pledge form used to reprogram a new personality.

POPF MODEL: The POPF Model (acronym for Professional contact-On television-Professional contact-Feedback) is derived from the PERF Model for TV-linked psychotherapy. (See PERF Model.) The POPF Model can be applied not only to psychotherapeutic sessions but also to any learning situation, whether between doctor-patient, counselor-client, teacher-student, family lawyer-family members, when personal contact can be meaningfully amplified by sandwiching TV tapes between personal contacts with proper professional feedback. In other words, to give a concrete example, the psychiatrist can see the patient in person at the beginning of a TV-linked therapeutic session, to elicit the chief complaint or problems; then the patient is left with the same psychiatrist on TV to amplify the therapeutic information. At the completion of a half-hour TV treatment program, the psychiatrist returns for further personal contact and relevant transaction regarding future appointments, medication, testing, etc., so that the loop is completed with feedback. The POPF Model can also be applied to couch therapy, family

THE PERF MODEL

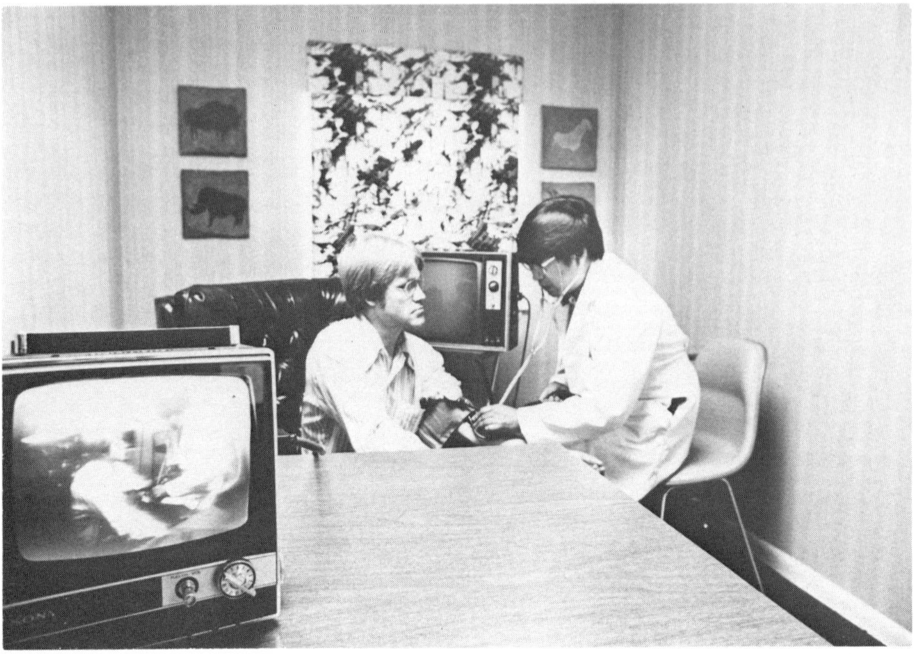

1. Patient is examined and analyzed. (*P*rogram).

2. Couch therapy is used to modify patient's memory (*E*dit)

3. A healthier personality is developed (*R*eprogram).

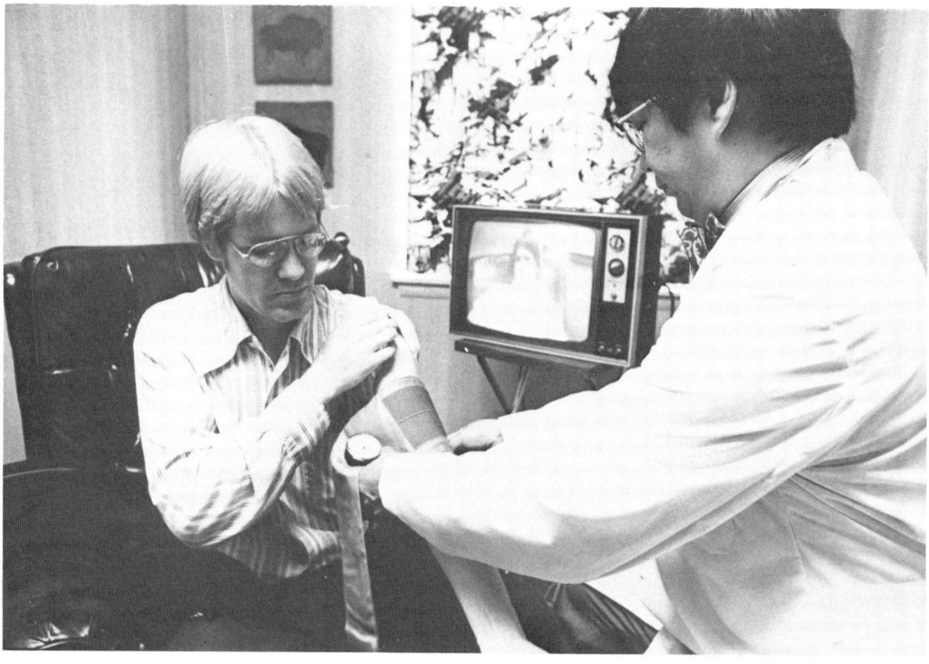

4. Patient views his therapeutic program (*F*eedback).

therapy, interview psychotherapy, hypnotherapy, etc. The general implication of the POPF Model is that one physician can become two physicians, or more, so that it may be an answer to the shortage of medical manpower, whether in the advanced industrialized countries like the United States and the Soviet Union, or in the nations in Asia, Africa, or South America. (See photographs which illustrate the POPF Model.)

POSTLYTIC: After lysis. After loosening of consciousness or memory in ELT. The postlytic step refers to the immediate re-programming of a new personality with a new name. For example: the prelytic step brings out the abnormal memories of an undesirable personality, called Anna, who, after undergoing electrolysis, is reprogrammed as Betty, a more desirable personality.

PRELYTIC: Before lysis. In psychosynthesis, the personality is viewed as a time-series of electrochemic self-creating or cybernetic pulses, which can be activated by free association, verbal stimuli or videotapes before lysis. The prelytic step of ELT refers to reactivation of memories and emotions before the actual application of electrical current for loosening and erasure of consciousness and memory.

PRENAME: The name of the pre-personality. For example: Anna is a person who undergoes electrolysis and is transformed into another personality, Betty. Anna is the prename of the pre-personality (prelytic personality).

PRIMARY TRANSNOMATION: The first therapeutic name change, usually the first name. E.g., Lisa is changed to Kathy. If Kathy changes her name again with a second series of ELT and transpersonation, then this transnomation, say from Kathy to Judy, is called Secondary Transnomation. Secondary Transnomation is rare these days (1970) and tertiary, etc., can be postulated.

PRINCIPLES OF PSYCHOSYNTHESIS: There are three basic principles of psychosynthesis and one universal law. For definition, see: (1) Most-similarity, Principle of; (2) Contradictions-in-identity, Principle of; and (3) Dynamic Thresholds, Principle of. See also: Universal Truth, Law of.

PROGRAMMABLE: The postlytic state in which the patient or synthesant is more receptive to master-program information.

PROGRAMMER: A person who feeds information into another person or systems in order to modify the memory of the system under programming.

PRO-PERSONALITY: The personality the individual aspires to become. For example, a violent personality may wish to achieve a peaceful personality. The peaceful personality is the pro-personality.

PSYCHOLYSIS: The process of loosening or erasing memory or consciousness. The loosening of mind for the purpose of personality transformation.

PSYCHOSCHISIS: Psychoschisis is the tendency to split off, to break away, and to separate from a unified, harmonious system of ideas. This process of separation, splitting or alienation is psychoschisis. Psychoschisis is a tendency for paranoid self-protection. Psychoschisis leads to sterile prejudices in the individual and destructive nationalisms in the world. Psychoschisis promotes narrow interests for individuals or small groups, but not to advance the interests of the health and well-being of the world community.

Those individuals who practice psychoschisis adjust their thresholds of perception by excluding others, by segregating themselves from a greater humanity.

PSYCHOSYNTHESIS: The term psychosynthesis means formation and transformation of personality by uniting all sub-personalities and fusing various techniques and theories for higher personality development, based on a unified theory of body-mind equivalence.

This unified definition has two basic origins: (1) Classical Psychosynthesis and (2) Cybernetic Psychosynthesis.

(1) Classical Psychosynthesis, as expounded by C. G. Jung, G. W. Henry, R. Assagioli, O. Diethem, S. Zweig, et al. The classical approach is an extension of psychoanalysis of Sigmund Freud. However, Freud had stated correctly, in 1924, that "I cannot imagine, however, that any new task for us is to be found in this psychosynthesis. If I were to permit myself to be honest and uncivil, I should say it was nothing but a meaningless phrase." (Freud, S. *Collected* Papers, vol. 2, tr. by Riviere, J., Leonard and Virginia Woolf and The Institute of Psychoanalysis, London, 1924–25.)

Freud's assessment and judgment were shared by Assagioli, as: "At this stage, since we do not have scientific instruments which enable us to measure these energies directly, we still have to rely on essentially a phenomenological position, in the sense of insisting on the experience itself, and hoping that sooner or later—maybe not in the lifetime of the author, science will attack this problem on a rigorous 'energy' basis." (Assagioli, R. *Psychosynthehesis*. Hobbs, Dorman & Company, Inc., New York, 1965, p. 194). Today (1970) the basic technology to develop the science of psychosynthesis is here. There are many tasks before us to develop in the new science of psychosynthesis based on cybernetics, information theory, television technology, etc. This is the cybernetic origin, according to the unified model of cybernetic psychosynthesis (Tien, 1969).

(2) Cybernetic Psychosynthesis. With the psychosynthesis model, Tien has fused the eclectic elements of modern psychiatry into a multiple TV-linked feedback system, composed of interview psychotherapy, couch analysis, family therapy, drug therapy, electrolytic therapy (ELT), group therapy, and super-group therapy as on-going dynamics of mental health mini-community of 50 families. In this system, the psychiatrist is also being transformed into a family psychiatrist, who could potentially work simultaneously with 100 families or more as the psychosynthesist of a mental health mini-community.

Psychosynthesis is now the telefusion of modern eclectic psychiatry. Psychosynthesis is a TV-linked cybernetic system of psychotherapy. Psychosynthesis is a unified theory of energy transformation. In theory, psychosynthesis is based on the unified concept of personality, transformation of brain-mind equivalence in the same sense of the mass-energy equation of Einstein. In practice, psychosynthesis is the cybernetic union of the psychoanalysis of Freud, behavior therapy of Pavlov, electrotherapy of Cerletti, and information theory of Shannon linked by modern television technology as a cybernetic process of Wiener. And so, the psychosynthesis model unites all schools of psychoanalysis, psychotherapy, and electrochemotherapy into a cybernetic system with a goal-orientation of the individual seeking his identity in the community.

In modern psychosynthesis, the term implies and includes the aspiration of the classical psychosynthesists (e.g., Jung, Assagioli), as the ancient atomists (e.g., Democritus, Lucretius) aspired for the understanding of matter in physics. And these ancient aspirations are being realized by modern technology. In the broadest sense, psychosynthesis includes the meaningful union of all contradictory ideals, ideas and techniques of modern eclectic psychiatry and psychology.

PSYCHOSYNTHESIST: A professional psychiatrist who practices a unified system of psychotherapy fusing classical analysis, synthesis, chemotherapy, electrolytic therapy, family therapy, group therapy, and community psychiatry with television as the storage, amplification and feedback subsystem.

PSYCHOSYNTHETIC: Pertaining to psychosynthesis. See Psychosynthesis, Mind Fusion.

R

REACTIVATION: The procedure used before electrolysis for the purpose of abreacting the patient or synthesant in order to loosen and erase memory and consciousness.

RESET: To return the matrix of zeros and ones all to zero as in the electrolytic process of psychosynthesis.

S

SCHISIS: See Psychoschisis.

SECOND PERSONALITY: The secondary personality, the undesirable personality or the prepersonality, which the individual attempts to shrink or eliminate from his conscious behavior, feeling and thoughts. Also, the competing personality, the rival personality to the controlling personality.

SETBACK: The reversion of the new personality to the previous personality, including the person's spontaneous use and social acceptance of his pre-name and pre-personality.

SOCIOSYNTHESIS: Society and combining parts to form a higher unity. Sociosynthesis is an extended concept of psychosynthesis. Sociosynthesis is understood in the restricted sense of combining two or more therapy groups into a super-group or a mental health community. The coffee shop of psychosynthesis is an expression of sociosynthesis. (See Coffee Shop.)

STATIC MEMORY: Stationed records in the brain. Stationary storage of electrochemical signs, which can be retrieved under proper stimulation. Static memory is genetic codes or experential signals. The genetic memory is permanent, whereas experential memory is semi-permanent. It is the semi-permanent memory, which is culture and language.

STONE MONKEY SYNDROME: The existential neurosis, probably universal to mankind. The syndrome is characterized by:

(1) physical fitness and material affluence;
(2) loss of purpose in life;
(3) boredom and restlessness;
(4) depression and suicidal ideation;
(5) the fear of death.

The Stone Monkey Syndrome is derived from Chinese folklore in the story of Wu Chengen's Xih Yyouzl Jiiyn (Journey to the West), which tells the aspiration of a stone monkey in search for immortality as a cure for his existential neurosis and fear of death.

Stone Monkey developed depression when he was surrounded by wealth and power but was unable to explain his "blues." His blues were due to the loss of purpose in life when he no longer had to struggle to hunt for food and to secure shelter. With loss of challenge in life, he lost reason for living; boredom and restlessness resulted. With boredom and restlessness, he found life to be empty, repetitious, depressing to the extent death would be preferable, even by his own hand. The idea of suicide scared him, so his first conscious symptom was the fear of death. The search for an antidote led him to his quest for immortality. This started his epic journey to the West. (See *Journey of the Stone Monkey* in modern transnarration by Tiann Honng Wenn.)

STRUGGLE: The process of actively seeking a new personality.

SUB-PERSONALITY: The personality which is not in dominant control of the person. The sub-personality may emerge either episodically as the executive personality or become the dominant personality. In psychosynthesis theory, a person has infinite personalities potentially, but the main feedback cycle of a most-similar pattern is the dominant personality, whereas the sub-cycles are the sub-personalities.

SYNTHESANT: One who is undergoing psychosynthesis.

SYNTHESIST: (1) One who subscribes to the union of all contradictory ideals, ideas, and practices of mankind. For example: A synthesist believes that both theist and atheist are correct; that both psychogenesist and organicist are correct; that both idealist and materialist are correct, etc., and that the higher truth is the union of all contradictions. A synthesist believes that all statements are true and truth is defined by threshold. He subscribes to the Law of Universal Truth.

(2) A short term for Psychosynthesist. (See also in Glossary.)

SYSTEM FUNCTION: Input-output relationship of a collection of elements with a unified information flow.

T

TELEFUSION: The use of two-way CCTV in videotherapy and videology.
TELESYNTHESIS: TST OR Telesynthesis therapy is telefusion couch or vis-à-vis therapy at a greater distance, for example, between two cities of states. TST may be used when the patient lives in a different city or when the therapist moves away to another state or even country.
TV-BOOSTER SESSION: A therapeutic television session whose aim is to increase doctor-to-patient communication by the use of preprogrammed TV tapes on all diagnostic, therapeutic, and medical regimes as a booster to person-to-person contact, with the doctor's presence on the TV monitor. During the same visit, the physician can personally provide simultaneous medical care to other family members or other families, e.g., a mother may be watching her family physician in a TV booster session explaining problems of pregnancy for unwed teenage girls, while her daughter is being examined for probable premarital pregnancy.
TV-COUCH SESSION: An analytic session being recorded on videotape for the purpose of playback, as used in feedback TV-therapy or reactivation in the prelytic step of electrolytic therapy.
TV-TAPE: Television tape. TV-tape is often referred to as videotape, but videotape is an incomplete description, since TV-tape contains an audio signal as a rule. In the psychosynthetic practice of psychiatry, TV-tapes form an integral part of the information storage and communication system.
THRESHOLD: The threshold T is defined as a preset energy level; when exceeded, the pattern recognition system, biological or computer, will discharge, otherwise not.

PRI (f, g) = 1, if $\phi_{fg}(\tau p)/[\phi_{ff}(0).\ \phi_{gg}(0)]^{\frac{1}{2}} \geq T$
and
PRI(f, g) = O, if $\phi_{fg}(\tau_p)/[\phi_{ff}(0).\ \phi_{gg}(0)]^{\frac{1}{2}} < T$,
where $\phi_{fg}(\tau_p)$ is the peak cross-correlation of two wave patterns, f and g, and $\phi_{ff}(0)$ and $\phi_{gg}(0)$ are their respective auto-correlations.

TOTE MODEL: The TOTE Model (acronym for Test-Operate-Test-Exit), first described by Miller, Galanter and Pribram, is derived from Wiener's cybernetic model. The TOTE Model is related to the pattern recognition by identity theory. (See Pattern Recognition by Identity Theory.) The TOTE Model can be seen as one possible mechanism specifying the interaction between "the two fundamental aspects of the pattern recognition system—system function and information." The functioning of the feedback loop is contingent upon the outcome of the tests which occur in the Test phase of the TOTE units. The TOTE Model is the primary cybernetic model, which can be applied to therapy, or counseling, or teaching in practical TV-linked systems of PERF Model and POPF Model. (See also PERF Model; POPF Model.)

TRANSNOMATION: The therapeutic name change. The name change can be either first name or last name. For example, an alcoholic personality, whose first name is Fred, is given a new name, John, in order to form a more desirable and more reputable identity and associations.

TWO-PERSONALITY THEORY: The two-personality theory states that every personality has at least two personalities. The theory is based on the following conceptual postulates:

(1) Consciousness is a single scintillating frame of a self-creating time-series in the cortex of the brain;

(2) Personality is the time-series of scintillating frames of the conscious neuron pattern; and

(3) Every time-series may be divided into two time-series.

If these postulates are accepted, then every personality can be divided into two personalities, which then could be divided into four personalities and four into eight, and so forth, *ad infinitum*. Every personality may then be divided into infinite personalities.

In the practice of psychosynthesis, the person is usually analyzed into two personalities. For example: Mrs. Anna Betty Oberman. If we let the surname Mrs. Oberman represent the physical person, then Mrs. Oberman has two personalities: Anna and Betty. If Anna is analyzed as the undesirable personality, then Betty, the desirable one, is emerging as Mrs. Oberman struggles to achieve the synthesis of her higher personality, i.e., Betty.

U

UNIVERSAL TRUTH, THE LAW OF: The Law of Universal Truth states that: Every statement is true, depending only on the threshold of acceptance.

Conversely, every statement is false, depending only on the threshold of rejection.

In psychosynthesis, the absolute value to determine truth is the absolute value of Life. That which leads to Life is true and that which leads to death is false.

V

VIDEOIST: (L. *VIDERE,* TO SEE,: VISION) A person who uses television technology in his occupation or profession.

VIDEOLOGIST: (*Video*, Vision; e.g., television + -logist) A physician who uses television technology extensively in diagnosing, treating and preventing diseases, while education and promotion of physical and mental health are being practiced through videology. (See Videology.)

VIDEOLOGY: (*Video*, Vision + -logy) The science and art of using television technology, especially in psychiatry and medicine, for the diagnosis and therapy of diseases and the rehabilitation and prevention of disorders, and the promotion of physical and mental health. This is analogous to radiology in modern medicine.

VIDEOSYNTHESIS: The application of psychosynthesis with videotherapy and telefusion psychotherapy.

VIDEOTAPE: TV-tape or Tevetape, a magnetic plastic tape used to record both television images and sound for broadcasting and feedback.

VIDEOTHERAPY: The use of CCTV as both passive and active elements in the treatment procedures of psychoanalysis and psychosynthesis for information transfer with free association and free communication techniques.

APPENDICES

Appendix A

PATIENT CONTACT FORM

NAME________________
ADDRESS______________
________ZIP____
PHONE NO.__________

RESPONSIBLE RELATIVE
Name________________
Relationship________
Occupation (or mate's)________
Address________
Phone No.________

FATHER'S EDUCATION________
Age__Occupation________
If deceased:
Cause________Age______

MOTHER'S EDUCATION________
Age__Occupation________
If deceased:
Cause________Age______

IDEOLOGY OR RELIGION
Patient________Spouse________
Father________Mother________

AGE__RACE__SEX__MARITAL__S__M__SEP
Birthdate____Place____
Patient's Education________
(Name of School)
Occupation__________

PATIENT'S SPOUSE
Name________Education______
Occupation________Age______
Patient living with________
Children's names & ages________
____

EMPLOYER OF PATIENT OR HEAD OF FAMILY:
____
Length of stay at job______
FAMILY'S MEDICAL INSURANCE:
Name__________
No.__________
PATIENT IN TOWN____Last Address______
Arrived via__________
Comment__________

Permission and consent are hereby granted to giving and receiving information to and from the following; and to prescribe procedures with information and intelligent participation of the undersigned in receiving service of family psychiatry and medicine with videotherapy.

REFERRED BY________________ ADDRESS________________
FAMILY PHYSICIAN________________ ADDRESS________________
OTHER PHYSICIANS OR PSYCHIATRISTS________________
OTHER THERAPISTS OF COUNSELORS________________
Typed By________Date________ Signed

APPENDIX B

TV-MONITORED EVALUATION FORMS

NAME______________ DATE________

1. INITIAL HISTORY AND EXAMINATION

0. INFORMANT: Are you answering or replying for yourself? Yes____ No____
   If No, for whom: Wife__ Husband__ Mother__ Father__ Other__

1. CHIEF COMPLAINT:
   A. What is the principal or the central reason you are coming to me?

   B. How long have you been concerned about this?
      Approximate date______________.

2. PRESENT DISTURBANCE: (Clarification of Chief Complaint)
   A. What concerns you most at this point?

   B. What other problems do you wish to bring up?

   C. What else really bothers you?

   D. What made you decide to seek professional opinion or help at this time?

   Explanation of Chief Complaint:
   A. When did you first begin to have this concern or problem?

   B. Perhaps you can't be sure of the problem, but what really bothers you?

   C. How does it bother you?

   D. Do you feel your problem is physical? Yes____ No____ Don't know____
      If Yes, describe briefly.

   E. Or is it emotional? Yes____ No____ Don't know____
      If Yes, describe briefly.

   F. Or is it mental? Yes____ No____ Don't know____
      If Yes, describe briefly.

   G. Or is it family? Yes____ No____ Don't know____
      If Yes, describe briefly.

   H. Or is it marital? Yes____ No____ Don't know____
      If Yes, describe briefly.

   I. Or is it a divorce problem? Yes____ No____ Don't know____
      If Yes, do you want to save your marriage? How?

   J. Have you had a similar situation before? Yes____ No____ Don't know____

K. Is it a school problem? Yes____ No____ Don't know____
If Yes, describe briefly.

L. Is it a sexual problem? Yes____ No____ Don't know____
If Yes, describe.

M. Is it a work or employment problem? Yes____ No____ Don't know____
If Yes, describe.

N. Do you have financial or money problems? Yes____ No____ Don't know____
If Yes, give figures, income, debts, etc.

O. Now that you have come to see me, what do you hope to achieve?

a. Do you want to improve your physical health? Yes____ No____ Don't know____
If Yes, name the physical problem.

b. Do you wish to have emotional stability? Yes____ No____ Don't know____
If Yes, state.

c. Do you wish to have peace of mind? Yes____ No____ Don't know____
If Yes, describe.

d. Do you wish to gain happiness? Yes____ No____ Don't know____
If Yes, clarify what is happiness to you.

e. Do you wish to improve family stability? Yes____ No____ Don't know____
If Yes, what areas?

f. Do you wish to improve your marriage? Yes____ No____ Don't know____
If Yes, what area?

g. Did you have any ideas of what kind of therapy you are seeking? Yes____ No____
If Yes, describe.

Or circle your preference. 1. Psychotherapy. 2. Couch Therapy Analysis.
3. Drug Therapy or Medicine. 4. Family Therapy. 5. Electrotherapy (i.e., ELT).
6. Group Therapy. 7. Personality Development. 8. Other______.

(If you need more space, use back of page.)

Appendix C

TV-MONITORED EVALUATION FORMS

NAME________________ DATE_______

3. PAST PROBLEMS

Be sure to answer every question. It will help to bring your past problems into view.

1. Have you had the same or similar problems before? Yes____ No____
   If Yes, give dates and describe.

2. Have you been to other psychiatrist? Yes____ No____
   If Yes, list the name(s) and dates.

3. What problems did you take to the previous doctor or doctors?

4. Why didn't you continue to work with these doctors? Give 3 reasons.
   1.
   2.
   3.

5. Do you see any connections between your present problems and your past?
   Yes____ No____ If Yes, describe.

6. Is there anything else you wish to say on your past or present problems?

(Please turn over for more space.)

Appendix D

TV-MONITORED EVALUATION FORMS

NAME________________ DATE_______

4. PREVIOUS TREATMENT

Be sure to answer every question. It will help to bring your problems into focus.

1. Have you ever had counseling? Yes____ No____ If Yes, from whom?

2. Have you ever been to a mental health clinic? Yes____ No____
   If Yes, where?

3. Have you ever been to a psychiatric hospital? Yes____ No____
   If Yes, where?

4. Have you ever had psychotherapy? Yes____ No____ If Yes, from whom?
   How many sessions?

5. Have you ever had nerve pills, tranquilizers? List them as follows:

Name of Drug	Shape	Color	How Often	What for
1.				
2.				
3.				
4.				
5.				
Which helped? (Check the number.)				

6. Have you ever had electro therapy? Yes_______ No_______
   If Yes, where?_______ When?_______
   How Many____ Describe_______

7. Is there anything else you wish to add about your previous or past treatment, medical or others?

(Please turn over for more space.)

Appendix E

TV-MONITORED EVALUATION FORMS
5. PERSONAL HISTORY

NAME_______________ DATE:_______

CONCEPTION: NORMAL: Yes__No__. Wanted: Yes__No__. Out of wedlock: Yes__No__Other__

BIRTH: Normal: Yes__No__. Full term: Yes__No__. Hospital delivery__Home delivery__

Any birth complications? Yes__No__. Birth injury: Yes__No__Describe______

NURSERY SCHOOL: Yes__No__. Describe adjustment________

GRADE SCHOOL: Began at age__. Describe adjustment________

HIGH SCHOOL: Completed through ____ year. Marks as a student: A B C D E (Circle correct letter)

Describe adjustment________________

COLLEGE: Completed through ____ year. Marks as a student: A B C D E (Circle correct letter)

Describe adjustment________________

DATING: Began at age ____. Steady dating at age ____. With whom? ________

How long?____. First important emotional relationship at what age? ____.

With whom? ________ How ended? ________

MARRIAGE: Age __. Type of wedding ________

Approval of both sides of family? Yes__ No__ Why? ________

If divorced, describe: ________________

EMPLOYMENT: First job at age ____. Type of job ________________

First important job at age ____. Type of job ________________

Last job at age ____. Type of job ________________

RELIGIOUS BACKGROUND: Brought up as ________ Practicing ________

POLITICAL VIEWS:

PHILOSOPHY:

Outlook on life:

SELF-DESCRIPTION: "I am

COMMENTS:

Appendix F

TV-MONITORED EVALUATION FORMS
6. FAMILY HISTORY

Please fill out the following form as completely as possible. List the first names of your father, mother, brothers, and sisters in chronological order and supply the following information about each:

NAME	AGE	EDUCATION and OCCUPATION	MARITAL STATUS	LIVING WHERE (City)	LIFE ADJUSTMENT good, fair or poor? (Explain)	COMMENTS
Father: (natural)						
(Step or adopting)						
Mother: (natural)						
(Step or adopting)						
Siblings: (Brothers and sisters—if more than 5 use back of form.)						
1.						
2.						
3.						
4.						
5.						
Spouse: (Husband or wife)						
Children: (All children of present or previous marriages—if more than 5, use back of form.)						
1.						
2.						
3.						
4.						
5.						
Ex-spouses:						
1.						
2.						

Your name: ______________ Date: ________

Appendix G

TV-MONITORED EVALUATION FORMS
7. SYSTEMIC REVIEW

Name_______________ Age______ Date______
Complexion____ Body Build____ Wt. ____ Ht. ____ Eye Color____ Hair color____

1. Have you ever been seriously ill physically?
2. When was your last medical examination?______ Name of Physician
3. To what extent have you used alcoholic beverages?
   Do you use or have you ever used any narcotic drug?
   Have you taken treatment for drug or liquor habit?
4. Previous hospitalizations ______ Date ______ Physician
   Nature of hospitalizations
5. Previous operations ______ Physician
   Nature
6. Previous trauma or accident ______ Date ______ Physician
   Nature
7. *Allergic* to any drug, food or substances?
8. *Eyes*: Do you wear glasses? ______ What for?
   Any other impairment?
9. *Ears*: History of deafness _______ Discharge_____
   Date ______ Any other impairment
10. *Nose*: History of nose bleeds, if any
11. *Neck*: Thyroid trouble _______ Glandular trouble_____
    Any other difficulty
12. *Heart*: History of heart attack _______ High blood pressure ____
    Rheumatic heart disease _______ Angina (heart pains) ____
13. *Lungs*: History of TB _______ Pneumonia ____
14. *Stomach*: History of vomiting _______ Stomach-ache ____
15. *Intestines*: Diarrhea _______ Constipation ____
    Gall Bladder ____ Liver disease ____ Other ____
16. *Urinary System*: Any disease of kidney
    History of syphilis _______ Gonorrhea ____
17. Have you ever had any glandular diseases?
    Sugar diabetes _______ Hyperthyroidism ____
    Addison's Disease _______ Other
18. Have you ever had any nerve disorder?
    Headache ____ Head trauma ____ Convulsions ____
    Loss of consciousness _______ Other
19. Have any of your parents, brothers, sisters, or near relatives ever had diabetes, mental illness, or tuberculosis? Who. _______
    Which _______
20. How long has your present weight been stationary?
    What caused the change of your weight?
21. Are you on any medications? If Yes, describe

22. *If you are a woman*
Age of onset of menstruation ____ Date of last period ____
Frequency and regularity ____ Length of cycle ____
Duration ____ Severity ____ Dysmenorrhea (menstrual difficulty) ____
Menopause ____ When ____
Are you pregnant? If so, give duration
How many pregnancies have you had? ______ Abortions ______
Have you ever had any disorder of the female organs?

COMMENTS:

Appendix H

8. PSYCHIATRIC EXAMINATION (Part I)

NAME____ DATE____

Patient appeared in consultation with ________ appearance,
neat, untidy, disheveled, etc.

dressed ________. H— cooperativeness was ________
appropriately, etc. good, fair, poor, etc

H— general manner was one of ________.
mistrust, suspicion, antagonism, preoccupation, etc.

H— general activity showed ________.
retardation, hyperactivity, tics, etc.

Patient's stream of mental activity was

H— speech was ________.
voluble, flighty, underproductive, mute, circumstantial, etc.

H— thought was ________.
logical, relevant, coherent, blocking, etc.

Emotionally, patient appeared to be ________
elated, euphoric, depressed, fearful, anxious,

________. H— affect was ________
irritable, angry, unstable, apathetic, etc. appropriate, ambivalent,

____
deteriorated, etc.

The content of the patient's thought showed evidence of ________
phobias, obsessions,

________.
psychosomatic complaints, delusions, hallucinations, depersonalization, etc.

H— sensorium was ________ with ________ conscious-
normal, abnormal clear, confused, cloudy

ness. Orientation was normal to ________. Memory was ________
time, place, person normal,

____
confabulatory, poor retention

Intellectually, the patient was able to read, write and calculate ________
not at all, grade,

________ level of education. Intelligence appeared to be ________
high school, college average,

________.
below average, above average, superior, retarded, etc.

The patient's judgment was ________ and h— insight was
good, defective, impulsive, etc.

____
lacking perceptive, obtuse, etc.

COMMENTS:

This form is included as part of the patient's complete history, but it is filled in by the physician.

Appendix I

TV-MONITORED EVALUATION FORMS

NAME________________ DATE______

8. (Con't.) Trend and Content Questionnaire (Part 2)

1. What do you worry about most of the time?

2. What do you fear?

3. Anything makes you angry?

4. Do you have any strong dislikes? Hatred?

5. Does anything or anyone make you jealous? Envious?

6. Anything makes you particularly sad?

7. If I gave you 3 wishes, what 3 important wishes do you have? (list them)
   Wish #1.
   Wish #2.
   Wish #3.
   Why these wishes?

8. Besides wishes, do you have dreams at night? Describe one:

9. Any repetitive dream? Describe it:

10. Why do you think you are now seeking professional help?

11. What are some of your explanations of your problems?

12. What ways have you tried to solve these problems?

13. Anything you wish to add?

(If you need more space, use back of page.)

Appendix J

8a. NEUROLOGICAL EXAMINATION
(Check & Describe)

Name________________ Date________

Sex__ Marital__ Education______ Ocupation______ Race__
Wt__ Temp__ Pulse__ Respiration__ BP__ Handed__ Eyed__

Referring Reasons:

Present Illness (for details, see back of page):

Previous Illnesses____ 4. Hospitalizations____ 5. Operations____
Systemic Review. 6.1 ENT ( ) 6.2 Neck ( ) 6.3 Cardiovascular ( )
6.4 Pulmonary ( ) 6.5 Gastrointestinal ( ) 6.6 Genitourinary ( )
6.7 Musculoskeletal ( ) 6.8 Endocrine ( ) 6.9 Others ( )

Neurological History. 7.1 Headache ( ) 7.2 Seizures ( ) 7.3 Pain ( )
7.4 Vision ( ) 7.5 Loss of consciousness ( ) 7.6 Weakness ( )

Neurological examination. 8.0 Fundoscopic OD __( ) OS__( )
8.1 Cerebral: 8.11 Behavior ( ) 8.12 Mood ( ) 8.13 Orientation ( )
8.14 Intelligence ( ) 8.15 Thought content ( )
8.16 Language ( ) 8.17 Insight ( ) 8.18 Others ( )
8.2 Cerebellar: 8.21 Gait ( ) 8.22 Romberg ( ) 8.23 Finger-to-nose ( )
8.24 Heel-to-shin ( ) 8.25 Adiadochokinesis ( )
8.26 Finger tapping ( ) 8.27 Rebound phenomenon ( )
8.3 Cranial nerves: 8.31 Olfactory ( ) 8.32 Optic ( ) 8.33 III, IV, VI ( )
8.34 Trigeminal ( ) 8.35 Facial ( ) 8.36 Auditory ( )
8.37 Glossopharyngeal ( ) 8.38 Vagus ( )
8.39 Spinoaccessory ( ) 8.39. Hypoglossal ( )
8.4 Sensory: 8.41 Pain ( ) 8.42 Temperature ( ) 8.43 Touch ( )
8.44 2-point tactile ( ) 8.45 Vibratory ( )
8.46 Positional ( ) 8.47 Stereognosis ( ) 8.48 Extinction ( )
8.5 Motor: 8.51 Neck ( ) 8.52 Arms ( ) 8.53 Trunk ( ) 8.54 Legs ( )
8.55 Atrophy ( ) 8.56 Hypertrophy ( ) 8.57 Spasm ( )
8.58 Fibrillation ( ) 8.59 Tonus ( ) 8.59.1 Others ( )
8.6 Reflexes: 8.61 Jaw Jerk ( ) 8.62 Biceps ( ) 8.63 Triceps ( )
8.64 Knee ( ) 8.65 Achilles ( ) 8.66 Abdominal ( )
8.67 Cremasteric ( ) 8.68 Babinski ( ) 8.69 Others ( )

Description of findings:

Tentative Diagnosis:
Laboratory Examinations Required: 11.1 EKG ( ) 11.2 EEG ( ) 11.3 BMR ( )
11.4 Spinal ( ) 11.5 Serology ( ) 11.6 X-Ray ( )
11.7 Urinalysis ( ) 11.8 Blood Analysis ( )
ll.9 Endocrine ( )
Recommendations:

Appendix K

8b. PHYSICAL EXAMINATION

Name________________ Date________

1. General:
2. B.P. Wt.
3. HEAD: ENT:
   EYES:
   NOSE:
   THROAT:
   MOUTH:
4. NECK:
5. CHEST:
6. HEART:
7. ABDOMEN:
8. GENITALIA:
9. RECTAL:
10. EXTREMITIES:
11. SPECIAL FINDINGS:

________________________ M.D.

Appendix L
D. I. M.
Depression Inventory Measure

Name________________ Date________

Read through these questions and *circle* the correct answer (a,b,c, or d) that describes you best.
Do not omit any items. It is very important that you should be quite frank.

1. Your mood is:
   (a) Happy
   (b) Sad
   (c) Very sad
   (d) Unbearably sad

2. Your outlook on life is:
   (a) Hopeless
   (b) Very discouraged
   (c) Discouraged
   (d) Optimistic

3. Your life is:
   (a) Not a failure
   (b) A partial failure
   (c) A failure
   (d) A complete failure

4. Your bowel movement is:
   (a) Regular
   (b) A little constipated
   (c) Rather constipated
   (d) Very constipated

5. Your conscience is:
   (a) Free
   (b) Guilty at times
   (c) Guilty most of the time
   (d) Burdened with guilt all the time

6. You regard yourself with:
   (a) Average praise
   (b) Need for correction
   (c) Deserving to be punished
   (d) Wishing to be punished

7. You feel:
   (a) Hatred for yourself
   (b) Disgusted with yourself
   (c) Disappointed with yourself
   (d) Satisfied with yourself

8. You judge yourself:
   (a) Fairly
   (b) Critically
   (c) Severely for many mistakes
   (d) Severely for everything

9. You wish yourself:
   (a) Life and well-being
   (b) With thought of harming yourself
   (c) With plans to commit suicide
   (d) With a definite plan to kill yourself

10. You cry or weep:
   (a) No more than usual
   (b) More than usual
   (c) All the time
   (d) Tears won't come anymore

11. Your irritability is:
   (a) No more than before
   (b) Easily set off
   (c) There all the time
   (d) Completely spent and dead

12. Your interest in people is:
   (a) Good as usual
   (b) Decreased
   (c) Greatly decreased
   (d) So little you don't care

13. You can make decisions:
   (a) No longer
   (b) Now only with help
   (c) Somewhat unsurely
   (d) As easily as before

14. Your appearance has:
   (a) Become repulsive
   (b) Become ugly
   (c) Changed for the worse
   (d) Remained the same

15. You can work:
    (a) Not at all
    (b) Only by pushing yourself
    (c) With difficulty to start
    (d) As well as usual

16. Your tiredness:
    (a) Comes with a day's work
    (b) Comes very easily
    (c) Comes with doing anything
    (d) Comes with doing nothing at all

17. You sleep:
    (a) Less than 5 hours a night
    (b) And wake up 1–2 hours earlier in the A.M.
    (c) And wake up unrefreshed
    (d) As well as usual

18. Your appetite is:
    (a) As good as ever
    (b) Not as good
    (c) Quite poor
    (d) Completely gone

19. You now feel:
    (a) Loved and liked as usual
    (b) Not as well loved and liked
    (c) Very few people love and like you
    (d) No one loves you

20. You worry:
    (a) Little about your physical health
    (b) About your aches and pains often
    (c) About your body all the time
    (d) That something is terribly wrong with your body

21. Your sex life is:
    (a) Completely gone
    (b) Of little interest
    (c) Less
    (d) Normal

22. Your life is:
    (a) Meaningless and senseless
    (b) Of little meaning
    (c) Losing some meaning
    (d) As meaningful as ever

23. Your thinking is:
    (a) As fast as ever
    (b) Somewhat slower
    (c) Rather slowed
    (d) Very slowed and difficult

24. Your movement is:
    (a) As quick as ever
    (b) Somewhat slower
    (c) Rather slowed
    (d) Very slowed and difficult

25. You feel:
    (a) No depression
    (b) Mild depression
    (c) Moderate depression
    (d) Severe depression

Appendix M

Name________________ Date________

MODIFIED MEDICAL SCHEDULE (ANXIETY SCALE)

Read through these questions and circle the correct answer, that describes you best, either "yes" or "no." Do not omit any items. It is important that you should be quite frank.

0. Are you going to be quite frank & honest in answering the questions? Yes No.
1. Do you have dizzy spells? Yes No
2. Do you get palpitations, thumping or racing of your heart? Yes No
3. Did you ever have a nervous breakdown? Yes No
4. Have you ever told a lie? Yes No
5. Did you often get "stage fright" or "audience fear" in your life? Yes No
6. Do you find it difficult to talk with strangers? Yes No
7. Have you ever made mistakes in your work? Yes No
8. Have you ever been knocked unconscious for 2 hours or more? Yes No
9. Do you worry too long over humiliating or embarrassing experiences? Yes No
10. Do you consider yourself rather a nervous person? Yes No
11. Are your feelings easily hurt? Yes No
12. Have you ever cut your finger or injured yourself in any way? Yes No
13. Are you subject to attacks of shaking or trembling? Yes No
14. Are you an irritable person? Yes No
15. Do ideas run through your head so that you cannot sleep? Yes No
16. Do you worry over possible misfortunes or bad things to happen? Yes No
17. Do you ever swear? Yes No
18. Do you sometimes feel happy, sometimes depressed, not knowing why? Yes No
19. Do you daydream a lot? Yes No
20. Do you seem to be less lively than others your age? Yes No
21. Have you ever criticized someone? Yes No
22. Do you have nightmares? Yes No
23. Do you worry about your health? Yes No
24. Have you sometimes walked in your sleep? Yes No
25. Do you sweat a great deal without exercise? Yes No
26. Do you find it difficult to make friends? Yes No
27. Do you ever make mistakes in your speech? Yes No
28. Are you touchy on various subjects? Yes No
29. Do you often feel dissatisfied or discontented? Yes No
30. Is it possible you give a false answer at times? Yes No
31. Do you often feel uncomfortable in the presence of your superior? Yes No
32. Do you suffer from sleeplessness? Yes No

33. Do you ever get short of breath without having done heavy work? Yes No
34. Are you ever forgetful? Yes No
35. Do you suffer from nerves? Yes No
36. Are you troubled with aches and pains? Yes No
37. Do you get nervous in places such as elevators, trains, or closed spaces? Yes No
38. Do you suffer from attacks of diarrhea or loose stool? Yes No
39. Are you lacking in self-confidence? Yes No
40. Are you troubled with feelings of not being good enough? Yes No
41. Have you ever been off work or school through sickness a good deal? Yes No
42. Have you ever been troubled by a stammer or stutter? Yes No
43. Do you usually keep in the background on social events or parties? Yes No
44. Are you rather shy or withdrawn? Yes No
45. Have you ever been angry with your mother or a close relative? Yes No
46. Do you sometimes get a pain over your heart? Yes No
47. Does your mind often wander badly, losing track of what you are doing? Yes No
48. Do you often feel just miserable? Yes No
49. Do you suffer from severe headaches? Yes No
50. Do you sometimes sneeze or cough in the presence of others? Yes No

Appendix N

P. P. S.
Psychiatric Prognostic Scale

Date________ Name________________

INSTRUCTIONS: Simply circle the answer describing your situation best as given within the brackets—choose one of the alternatives.

1. YOUR AGE IS (Under 20, 30, 40, 50, above 50) YEARS.
2. YOU HAVE HAD EMOTIONAL DISTURBANCES FOR (Less than 1, 3, 5, 10, over 10) YEARS.
3. YOU FINISHED (College, High School, 8th grade, elementary, none).
4. YOUR DISTURBANCE BOTHERS YOU (Slightly, some, so-so, much very much.
5. YOU HAVE HOW MANY BLOOD RELATIVES WITH MENTAL ILLNESS (None, 1, 2, 3, over 3).
6. YOUR CHILDHOOD WAS (Very happy, happy, so-so, unhappy, horrible).
7. YOU WANT TREATMENT (Very much, much, so-so, some, not at all).
8. WHAT DO YOU FEEL FOR YOUR FATHER? (love, like, mixed, dislike, hate).
9. WHAT DO YOU FEEL FOR YOUR MOTHER? (love, like, mixed, dislike, hate).
10. YOU LIKE PEOPLE (very much, much, so-so, some, not at all).
11. HAVING DONE WRONG, HOW GUILTY DO YOU FEEL? (quite, strongly, terribly, not much, not at all).
12. YOU GET WHAT YOU REALLY WANT (almost always, often, sometimes, seldom, never).
13. YOU HAVE BEEN HOSPITALIZED FOR PHYSICAL ILLNESS (not once, 1, 2, 3, more than 3 times).
14. YOU HAVE BEEN HOSPITALIZED FOR EMOTIONAL OR MENTAL ILLNESS (not once, 1, 2, 3, more than 3 times).
15. YOU HAVE BEEN IN JAIL OR PRISON (not once, 1, 2, 3, more than 3 times).
16. YOU HAVE BEEN DIVORCED (not once, 1, 2, 3, more than 3 times).
17. YOU ARE (rich, well-to-do, adequate financially, poor, in debt).
18. YOU HAVE CHANGED JOBS WITHIN THIS YEAR (not once, 1, 2, 3, more than 3 times).
19. YOU HAVE CHANGED YOUR PLACE OF RESIDENCE WITHIN THIS YEAR (not once, 1, 2, 3, more than 3 times).

20. YOU FINISH WHAT YOU PLAN TO DO (almost always, often, sometimes, rarely, never).
21. BEFORE YOUR PRESENT PROBLEM YOU HAD WHAT KIND OF A LIFE ADJUSTMENT? (excellent, good, fair, poor, horrible).
22. YOU ENJOY FOOD (very much, much, so-so, some, not at all).
23. YOU HAVE HAD HOW MANY PSYCHIATRISTS PREVIOUSLY? (none, 1, 2, 3, more than 3).
24. YOU MAKE FRIENDS (very easily, easily, so-so, with difficulty, rarely).
25. YOU FEEL PSYCHIATRIC TREATMENT WILL HELP YOU (greatly, much, some, just a little, not much).

COMMENTS OR EXPLANATIONS:

Appendix O

PQ: THE PROBLEM QUESTIONNAIRE

Name________________ Date________

INSTRUCTIONS: Answer the following questions frankly please. Write and check in detail. Answer all items. This is a helpful aid to bring your treatment into focus.

My most important problem today is:

Do you have any questions about:
(a) Your medicine or the treatment today? No__ Yes__
(b) If yes, describe:

Do you have any special physical problems this week? No__ Yes__
If yes, describe:

Do you have any special emotional problems this week: No__ Yes__
If yes, describe?

Do you have any new family problems this week? No__ Yes__
If yes, describe:

Do you anticipate or foresee any treatment problems next session, such as scheduling, planning a vacation, etc.? No__ Yes__
If yes, describe:

I would like to know more about: Diagnosis__ Prognosis__ Medicine__ Couch Sessions__ Family Sessions__ Marital Therapy__ Electrolytic Therapy (ELT)__
Group Therapy__ Telefusion Sessions__ TV-linked Sessions__
Other, describe

I understand MIP Therapeutic Fee Policy: Yes__ No__

The other problems on my mind are:

(If you need more space use back of page.)

Appendix P

NOTES ON TV-TREATMENT PROGRAMS

Name________________ Date________

(A) Subject of the program: (Give title or your own title.)

(B) Important points:

(C) Words I don't understand:

(D) Questions I would like to ask:

(E) Other comments:

(F) Please answer the following questions concerning the program: (Circle one.)

1. The program itself was: Excellent / Good / Fair / Poor

2. For me, the program was: Very helpful / Helpful / Somewhat Helpful/Not helpful

(G) We would like to know more about:

(If you need more space use back of page.)

Appendix Q

NOTES ON TELEFUSION COUCH FEEDBACK

NAME_______________ DATE_______

INSTRUCTIONS: Please take notes as detailed as you can. Give direct quotations when significant to you. Record dreams, fantasies, conflicts, and recurrent statements or repeated problems. *Use both sides of the page if possible*.

1. Please describe the beginning of the session:

2. What thoughts, feelings, and ideas were brought out?

3. The names of people, relatives, friends, teachers, etc., mentioned:

4. Write down any topic or ideas that are repeated in the session:

5. Write down unsolved problems, questions:

6. The areas I would like to explore in the future are:

7. The last thoughts on the tape were:

8. (Please circle your comment regarding the feedback):
   The feedback itself was: Excellent / Good / Fair / Poor
   For me, the feedback was: Very helpful / Helpful / Somewhat helpful / Not helpful

9. I would like to suggest some changes to help my treatment:

(If you need more space, use back of page.)

Appendix R

PROGRESS REPORT

INSTRUCTIONS: The following is an important form for you to complete, whether for yourself or for the member of your family directly concerned. I am interested in knowing what progress you feel we are making. Fill in the questions below and answer them as objectively as you can.

Please check the following:

1. *The symptoms and complaints for which we sought treatment originally are:*
( ) Much better ( ) Better ( ) No change ( ) Worse ( ) Much worse

2. *Our understanding of the condition is:*
( ) Excellent ( ) Good ( ) Fair ( ) Poor ( )Very Poor

3. *We would consider the progress to be:*
( ) Excellent ( ) Good ( ) Fair ( ) Poor ( ) Very Poor

4. *The following treatments have been helpful to us:*
Please number in order of helpfulness: A, B, C . . . etc., as applicable to you.
( ) Medicine ( ) Individual Session ( ) TV Session ( ) Electrotherapy (ELT)
( ) Family Session ( ) Couch Session ( ) Group Session ( ) Other (specify)
( ) Videotherapy ( ) Telefusion ( ) Booster Videotherapy

5. *We feel the present treatment should be:*
( ) Same ( ) More Often ( ) Less Often ( ) Other (specify).

6. *You have had some treatment from another doctor since your last visit:*
( ) Yes ( ) No. If *Yes*, please give details:

7. *Now, I (we) understand the MIP Therapeutic Fee Policy*: ( ) Yes ( ) No

8. *Today, I (we) choose to pay according to*: (Check one)
Plan A ( ) Plan B ( ) Plan C ( ) Plan D ( ) Plan E ( )
Explain:

Name________________ Date________

If you have any additional comments about your past treatment or present therapeutic plan, please use back of this page. Thank you.

Appendix S

CONSENT TO TAKING OF VIDEOTAPES

Patient ____ Date ___ City & State ____

The undersigned, I (or we) grant this consent as a free and voluntary contribution for and on my (or our) own treatment. And I (or we) acknowledge fully the fact that this recording will be used only for therapeutic purposes, medical education, and the advancement of family health.

I (or we) hereby consent to the taking of videotapes of the medical procedures which are scheduled to be done with me and my family from the above date to and throughout the duration while under treatment or receiving service from this clinic.

I (or we) hereby authorize our attending physician, whenever necessary, to admit his medical assistants and technicians who will participate in said video-taping.

SIGNATURES

Witness: ____	Signed: ____
____ ____	Relationship: ____
____ ____	Signed: ____
____ ____	Relationship: ____

BIBLIOGRAPHY

1. Affleck, D. C. "Two-Way TV in Group Psychotherapy." Coune. Med. Telev. Newsl. 2:2–3.
2. ——— "Two-Way TV in Group Therapy." Coun. Med. Tele. News 2–3.
3. Alger, I and Hogan, P. "The Impact of Videotape Recording on Involvement in Group Therapy." J. Psycho. Groups 2:50–56.
4. ———, and ——— "Use of Videotape Recordings in Conjoint Marital Therapy in Private Practice." Amer. J. Psychiatry, 123:1425–1430, 1967.
5. ———, and ——— "Enduring Effects of Videotape Playback Experience on Family and Marital Relationships." Amer. J. Orthopsychiatry 39:86–98.
6. ———, and ——— Therapeutic Use of Videotape Playback. J. Nerv. Ment. Dis. 148:430–436.
7. Alktre, A. S. and Brunse, A. J. Impact and Possible Casualty from Videotape Feedback in Marital Therapy. J. Consulting and Clinical Psych. 42:203–210.
8. Anderson, C. J., and Sainanto, H. K. "Use of Videotape Feedbacks as a Psychotherapeutic Medium Approach with Long-term Psychiatric Patients—A Pilot Study." Nurs. Res. 22:525–530.
9. Apostoles, E. F., Anderson, J., and Hoye, C.L. "The Use of Closed Circuit Television in Marital Therapy." J. Psych. Nurs., 12:10–16.
10. Bahnson, C. "Body and Self-Images Associated with Audio-Visual Self-Confrontation." J. Nerv. Ment. Dis., 148:262–180.
11. Bailey, K. G., and Sower, W. T. "Audiotape and Videotape Self-Confrontation in Psychiatry." Psych. Bull. 74:2:127–137.
12. Barlow, D. J., and Bruhn, J. G. "Role-Plays of Television—A New Teaching Technique. Nurs. Outlook 21:242–244.
13. Barnes, L. H., et al "Psychiatric Patients and Closed Circuit Television Teaching: A Study of Their Reactions." Brit. J. Med. Educ. 3:58–61.
14. Benschoter, R. A., Cecil, L., Wittson, C. L., and Ingham, C. G. "Teaching and Consultation by Television, I: Closed Circuit Collaboration," Ment. Hosp. 16:99–100.
15. ———, Eaton, M. T., and Smith, P. "Use of Video-Tape to Provide

Individual Instruction in Technics of Psychotherapy." J. Med. Ed. 40:1159–1161.

16. ——— "Multipurpose Television." Ann. N.Y. Acad. Sci. 142:471–478.
17. Berger, M. M., and Gallant, D. M. "The Use of Closed Circuit Television in Teaching of Group Psychotherapy, Psychosomatics." 6:16–18.
18. ——— "Therapist Insight Sharpened by Video Technique; Wide Use in Private Practice Predicted." Psychiat. Progr. (Eli Lilly) 3:2:3–5, 1968.
19. ———, Sherman, B., Spalding, J., and Westlake, R. "The Use of Videotape with Psychotherapy Groups in a Community Mental Health Service Program." Int. J. Group Psychother. 18:504–515.
20. ——— "Integrating Videotape into Private Practice." Voices, in press, 1970.
21. ——— "Videotape Techniques in Psychiatric Training and Treatment. New York, Brunner/Mazel, 1970.·
22. ——— "Self-Confrontation Through Video." Am. J. Psychoan. 31:48–58.
23. ——— "Distorted Video Images Quickly Elicit Repressed Material." Roche Report, 3:(9):1–3, 1973.
24. ——— "A Preliminary Report on Multiple Image Immediate Impact Video Self-Confrontation," Am. J. Psych. 130:304–306.
25. Bergman, A. C. "Videotape Self-confrontation of Schizophrenic Ego and Thought Processes." J. Consulting and Clinical Psych. 39(1):78–85.
26. Blancheri, R. L., and Merrill, L. R. "Television in Health Sciences Education: Camera Placement." Nurs. Res. 13:217–221.
27. Blount, H. P., and Pederson, A. M. "Effects of Video Playback of a Person on His Self-concept." Psychological Reports 26:667–670.
28. Bodin, A. M. "Videotape Applications in Training Family Therapists." J. Nerv. Ment. Dis. 148:251–261.
29. Brayton, D. "Postgraduate Education in Medicine by Television." Biomed. Sci. Instrum. 3:81–88.
30. Chandler, M. J., Grienspan, S., and Barenboum, C. "Judgments of Internationality in Response to Videotaped and Verbally Presented Moral Dilemmas—The Medium is the Message." Child Development, 44:315–320.
31. Coone, J. G., et al. "Role of the Classroom Instructor in a Televised Introductory Psychology Course." Psychol. Rep. 23:43–47.
32. Cornelison, F. S., and Tausig, T. M. "A Study of the Self-Image Experience Using Videotape at Delaware State Hospital." Del. Med. J. 36:229.
33. Cozean, C. E., and Reivich, R. "TV in Psychiatric Nursing Education: A Preliminary Report." Counc. Med. Telev. Newsl. 2:.
34. Czajoski, E. H. "The Use of Videotape Recordings to Facilitate the Group Therapy Process." Int. J. Group Psychother. 18:516–524.

35. Danet. B. N. "Self-Confrontation in Psychotherapy Reviewed: Videotape Playback as a Clinical and Research Tool." Amer. J. Psychother. 22:2:246–257.
36. ——— "Impact of Audiovisual Feedback on Group Psychotherapy." J. Cons. and Clinical Psych. 33:362–370.
37. ——— "Videotape Playback is a Therapeutic Device in Group Psychotherapy." Cit. J. of Group Psychotherapy 19(4):433–440.
38. Davies, T. S. "Closed-Circuit Television in a Mental Hospital." Brit. Med. J. 5318:1531–1532.
39. Dwyer, T. E. "Telepsychiatry—Psychiatric Consultation by Interactive Television." Am. J. Psychiatry 130:865–869.
40. Feristein, C., and Tamerin, J.: "Induced Intoxication and Videotape Feedback in Alcoholism Treatment." Quarterly J. Studies on Alcoholism. 33(2).
41. Forrest, D. V., Ryan, J. J., Galvin, R., and Merrital, H. H. "Through the Viewing Tube—Videocassette Psychiatry." Am. J. Psych. 131:90–94.
42. Furman, S., and Elighman, A.: "Video Feedback in Treating Hyperactive Children—A Preliminary Report." Am. J. Psychiatry 130:792–796.
43. Gant, H. M. "Studies of Closed-Circuit Television in Psychotherapy Yield Promising Results." Psychiat. News 3:8–9.
44. Gauson, F. F., and Raulings, F. T. "Using Videotapes to Treat Chronic Patients in Vicarious Therapy Groups." Hospital Community Psychology 25:277.
45. Gelso, C. J. "Effect of Audiorecording and Videorecording on Client Satisfaction and Self-expression." J. Consulting and Clinical Psych. 40:455–461
46. Geertsma, R. H., and Reivich, R. S. "Repetitive Self-Observation by Videotape Playback." J. Nerv. Ment. Dis. 141:29–41.
47. Gildfield, M. D., and Levey, R. "The Use of Television Videotape to Enhance the Therapeutic Value of Psychodrama." Am. J. Psychiatry 125:690–692.
48. Gioscia, V. "Notes on Video Therapy." Radical Software 2(4):1–4.
49. Gladfelter, J. "The Use of Video Tape Recording for Supervision of Group Psychotherapists."
50. Goldfield, M. D., and Levy, R. "The Use of Television Videotape to Enhance the Therapeutic Value of Psychodrama." Amer. J. Psychiat, 125:690–692.
51. Gould, J., and Smith, K. U. "Sensory Feedback Analysis of Stereotelevision Pursuit Tracking." J. Appl. Psychol. 48:152–160.
52. Green, J. H. et al. "Use of Videotape Recording and Live Closed-Circuit Television in Teaching Medical Students." Brit. J. Med. Educ. 1:135–143.
53. Griffin, G. J., Kinsinger, R. E., and Pitman, A. J. "Clinical Nursing Instruction and Closed Circuit TV." Nurs. Res. 13:196–204.
54. Gruenberg, P. B., Liston, E. H., and Wayne, G. J. "Intensive

Supervision of Psychotherapy with Videotape Recording." Amer. J. Psychother. 23:98–105.
55. Harris, J. J. "Television as an Educational Medium in Medicine: An Historical Purview." J. Med. Ed. 41:1–19.
56. Hirsh, H., and Freed, H. "Pattern Sensitization in Psychotherapy Supervision by Means of Video Tape Recording." Psychiat. Spect. 5:7–8.
57. Hogan, P., and Alger, I. "Impact of Videotape Recording on Insight in Group Therapy." Int. J. Group Psychother. 19:158–165.
58. Holmes, D. J. "Closed Circuit Television in Teaching Psychiatry." Univ. Mich. Med. Bull. 27:330–336.
59. Jaramillo, J. D. C. "Video-oriented Psychiatric Practice." Proceeding, First Regional Congress of Social Psy., Calif. 1977.
60. Kagan, N., Krathwohl, D. R., and Miller, R.: "Stimulated Recall in Therapy Using Videotapes: A Case Study." J. Counsel Psychol. 10:237–243.
61. Kornfeld, D. S., and Kolb, L. C. "The Use of Closed Circuit TV in the Teaching of Psychiatry." J. Nerv. Ment. Dis. 138:452–459.
62. Krystal, H. "Postgraduate Teaching in Psychiatry: Combining Television and FM-Radio with Seminars." Psychosomatics 3:1–4.
63. Kubie, L. S. "Some Aspects of the Significance to Psychoanalysis of the Exposure of a Patient to the Televised Audiovisual Reproduction of His Activities." J. Nerv. Ment. Dis. 148:301–309.
64. Lautch, H. "Video-tape Recording as an Aid to Behavior Therapy." Brit J. Psych. 117:207–208.
65. Maclay, D. T. "Teaching Psychiatry by Closed-Circuit Television." Brit J. Psychiat. 114:918–919.
66. Mann, J. "Vicarious Desensitization of Test Anxiety Through Observation of Videotaped Treatment." J. Consulting Psych. 19(1):107.
67. Martin, G. L., and Over, H. R. "Therapy by Television." Audiov. Commun. Review 4:119–130.
68. Marvit, R. C., Lind, J., and McClaughin, D. G. "Use of Videotape to Induce Attitude Change in Delinquent Adolescents." Am. J. Psych. 131:(9)317–319.
69. McDermott, J. F., and Harrison, S. I. "Some Considerations in the Use of Television during the Clinical Years." J. Med. Educ. 39:889.
70. McGuire, F. L., Moore, F. J., Harrison, C. A., and Riley, R. E. "The Efficiency of Television as Applied to the Use of Laboratory Demonstrations in Teaching." J. Med. Educ. 36:715–716.
71. McGuire, F. L., and Stigall, T. T. "Psychotherapy by Television: A Research Paradigm" Psychother.: Theory, Res., Pract. 3:159–162.
72. McNiff, A., and Cook, O.: "Video Art Therapy." Art Psychotherapy 2:55–63.
73. Meadow, R., and Hewitt, C. "Teaching Communications Skills With the Help of Actresses and Videotape Simulation." Br. J. Med. Ed. 6:317–319.

74. Melnick, J., and Tims, A. R. "Application of Videotape Equipment to Group Therapy." Int. J. Group Psychotherapy 24:10–16.
75. Michaux, M. H., Cohen, M. J., and Kurland, A. A. "Closed-Circuit Television in the Scientific Measurement of Psychopathology." Med. Biol. Illus. 13:49–57.
76. Miller, D. "The Effects of Immediate and Delayed Audio and Videotape Feedback in Group Counseling." Comparative Group Studies 1:47–77.
77. Millerson, G. "The Techniques of Television Production". New York, Hastings House Publishers, 1970.
78. Moore, F. J., Russell, S. C., Hubbard, O. E., Davis, C. R., Stuart, B. R., and Overall, J. E.: "Clinical Judgement: The Effectiveness of Television and Video Tape for the Symptomatic Evaluation of Psychiatric Patients." Submitted for Publication, 1977.
79. ———, Hanes, L. C., and Harrison, C. A. "Improved Television, Stereo and the Two-Person Interview." J. Med. Educ. 36:162–166.
80. ———, Chernell, E., West, M. J. "Television as a Therapeutic Tool." Arch. Gen. Psychiat. 12:217–222.
81. Moreno, J. L. "Television Videotape and Psychodrama." Amer. J. Psychiat. 125:1453–1454.
82. Onder, J. J. "The Use of Television in Psychiatric Treatment and Education." J. Soc. Motion Pict. Telev. Engrs. 77:1034–1037.
83. ———, "A Review of Television Formats for Students of Psychiatry." Health Sci. TV Bull. 6:1–6.
84. ———, Manual of Psychiatric Television. Ann Arbor, Michigan, Maynard House, 1970.
85. Pascal, G. R., Cottrell, T. B., and Baugh, J. R. "A Methodological Note on the Use of Video Tape in Group Psychotherapy with Juvenile Delinquents." Int. J. Group Psychother., 17:248–251.
86. Perlmitten, M., Loeb, G., and O'Hara, H. "Family Diagnosis and Therapy Using Video Tape Playback." Am. J. Orthopsychology 37:900–905.
87. Ramey, J. W. "The Physician's Responsibility to the Patient: Videotape Problem-Solving in a Self-Instructional Mode." In Lysaught, J. P., ed: Individualized Instruction in Medical Education. Rochester, N.Y., The Rochester Clearing House, 1968.
88. ———, "Teaching Medical Students by Videotape Simulation." J. Med. Educ. 43:55–59.
89. Rawls, J. R. et al. "Evaluation of Closed Circuit Television in Teaching Educational Psychology." Psychol. Rep. 22:1041–1044.
90. *Reivich, R. S.* and *Geertsma, R. H.* "Experience with Video Tape Self-Confrontation by Psychiatric Inpatients." J. Kamas Soc. 69-:39–44.
91. ———, et al. "Television and Psychiatry. Observation Media in Psychiatry: The Concept of a Learning Laboratory." J. Kansas Med. Soc. 70:101–104.
92. Resnik, H. L., Davison, W. T., Schuyler, D., and Christopher, D.

"Video Tape Confrontation After Attempted Suicide." Am. J. Psych., 130:46–463.
93. Resnikoff, A., Kagan, N., and Schauble, P. "Acceleration of Psychotherapy Through Stimulated Videotape Recall." Am. J. Psychotherapy 24:102–111.
94. Richter, H. E.: "Television Transmission of Psychoanalytic Interviews," 1–19, 1969.
95. Robinson, M., and Jacobs, A. "Focused Videotape Feedback and Behavior Change in Group Psychotherapy." Psychoatherapy; Theory, Research and Practice, 7(3): 169–172.
96. Rogers A. H. "Videotape Feedback in Group Psychotherapy." Psychother, Theory, Res., Pract. 5:1:37–40.
97. Rogus, A. "Videotape Feedback in Group Psychotherapy." Psychotherapy Res. Ther. Practice, 5:37–39, 1968.
98. Ruhe, D. S., Gundel, S., Laybourne, P. D., Forman, L. D., Jacobs, M., and Eaton, M. T. "Television in Teaching Psychiatry." J. Med. Educ. 35:916–927.
99. ———, and Gentry, K. C. "Ten Years of Television at the University of Kansas Medical Center." J. Kansas Med. Soc. 62:94–99.
100. Ryan, J. "Teaching and Consultation by Television: II. Teaching by Video-tape." Ment. Hosp., 16:101–104.
101. Rynearson, R., and Wilmer, H. "Video as a Medium in Psychotherapy." Current Psychiatric Therapies 10:83–87.
102. Schaefer, H. H., Sobell, M., and Sobell, L. "Twelve Month Follow-up of Hospitalized Alcoholics Given Self-Confrontation Experiences by Videotape." Behavior Therapy, 3:283–285.
103. Schiff, S. B., and Reivich, R. S. "Use of Television as an Aid to Psychotherapy Supervision." Arch. Gen. Psychiat. 10:84–88.
104. Sclare, A. B. et al. "The Use of Closed Circuit Television in Teaching Psychiatry to Medical Students." Brit. J. Med. Educ. 2:226–228.
105. Scott, M. "Television Hypnosis." JAMA 193:4:854–855.
106. Sethna, E. R., and Neal, C. D. "Television in Psychotherapy." Brit. Med. J. 5457:364.
107. Shostrom, E. L. "Witnessed Group Therapy on Commercial Television." Amer. Psychol. 23:207–209.
108. Skamia, H. J. Television and Society. New York, McGraw-Hill, 1965.
109. Smith, K. U., and Gould, J. "Sensory-Feedback Analysis of Behavior in Stereo-Televised Visual Fields." J. Appl. Psychol. 48:361–368.
110. ———, and Smith, T. J. "Systems Theory of Therapeutic and Rehabilitative Learning and Television." J. Nerv. Ment. Dis. 148:386–429.
111. Solow, C., Weiss, R., and Berger, B. "24-Hour Psychiatric Consultation Via TV." Am. J. Psych. 27:1684–1687.
112. Stoller, F. H. "Group Psychotherapy on Television. An Innovation

with Hospitalized Patients." Amer. Psychol. 22:158–162.

113. ———, "Focused Feedback with Videotape: Extending the Group's Functions." in Gazda, G. M., Ed: Basic Innovations in Group Psychotherapy and Counseling. Springfield, Ill., Charles C Thomas, 1968.

114. ———, "Use of Videotape (Focused Feedback) in Group Counseling and Group Therapy." J. Res. Develop. Educ. 1:30–44.

115. ———, "Videotape Feedback in the Group Setting." J. Nerv. Ment. Disc. 148:457–466.

116. Stroms, M. D. "Videotape and the Attribution Process Diversing Actors and Observers Points of View." J. Res. Soc. Psych. 27:165–175.

117. Suess, J. F. "Teaching Clinical Psychiatry with Closed Circuit Television and Videotape." J. Med. Educ. 41:483–488.

118. ———, "Self Confrontation of Videotaped Psychotherapy as a Teaching Device for Psychiatric Students." (Submitted for Publication, March 1969).

119. Tausig, T. N., and Shaeffer, S. "Self-Image Experience by Immediate Television Feedback: A Preliminary Report." Unpubl. paper.

120. Thompson, J. W., Nash, L. R., and Emmitt, T. "Closed-Circuit TV Used for Psychotherapy." Mod. Hosp. 195:60–61.

121. Tien, H. C. "Psychosynthesis: Television-Linked Cybernetic System of Psychotherapy." Given at the International Congress of Cybernetics, London, England, September 1969.

122. ———, "Psychosynthesis: TV-Cybernetic System of Psychotherapy." Given at the Third Annual Symposium, American Society for Cybernetics, Gaithersburg, Maryland, October 1969.

123. ———, "Psychosynthesis: TV-Linked Cybernetic System Psychotherapy." in Rose, J. ed.: Progress of Cybernetics, Vol. I." London, Gordon and Breach Science Publishers, Ltd. 1970.

124. ———, "Psychosynthesis: A TV-Cybernetic Hologram Model." in Wulfsohn Norman L. and Sances, Anthony Jr., eds.: the Nervous System and Electric Currents, Vol. 2. New York, Plenum Press, 1971.

125. ———, "Psychosynthesis: TV-Linked Practice of Psychosomatic Medicine." Given at the First International Collegium of the International College of Psychosomatic Medicine, Guadalajara, Mexico, December 1971.

126. ———, "The Science of Videology: TV-Monitored Evaluation Forms." World Journal of Psychosynthesis. Vol. 4; No. 3, March 1972, pp. 30–42.

127. ———, "Psychosynthesis with Television Technology." in Massorman, Jules, ed.: Current Psychiatric Therapies, Vol. 15. Grune and Stratton, New York, 1975.

128. ———, "Seminar on Videotechniques and Videotherapy." First Regional Congress of the International Association for Social Psychiatry, September 1977.

129. ———, "Videotechniques in Family Psychiatry." First Regional Congress of the International Association for Social Psychiatry, September 1977.
130. Toeman, Z. Audience Reactions to Therapeutic Films. Group Psychotherapy: A Symposium. New York, Brown House, 1945, pp 255, 259.
131. Torkelson, L. O., and Romano, M. T. "Self-Confrontation by Videotape? A Remedial Measure in Teaching Diagnostic Evaluation." JAMA 201:773–775.
132. Trethowan, W. H. "Teaching Psychiatry by Closed Circuit Television, Brit. J. Psychiat.," 114:517,522.
133. Trull, S. G. "California Medical Team Videotapes Sessions." Med. Tribune 10:.
134. Warren, F. Z. "Medical Television: An Evaluation." J. Med. Educ. 31:418.
135. Watts, J. C. "Closed-Circuit TV in Psychiatric Hospitals: An Aid to Nursing." Nurs. Times 60:15–17.
136. Wilhite, J. A. "A Manual on the Use of Video in Psychotherapy." Unpublished Master Thesis, Goddard College, Vermont, 1976.
137. Wilmer, H. A. "The Undisguished Camera in Videotape Psychiatric Teaching." Paper presented at the 7th Annual Meeting of the Council on Medical Television at the University of California Medical Center, September 1966.
138. ———, "Practical and Theoretical Aspects of Videotape Supervision in Psychiatry." J. Nerv. Ment. Dis. 145:123–130.
139. ———, "A Role of Psychiatrist in Consultation and Some Observations on Videotape Learning." Psychosomatics 8:193–195.
140. ———, "Television: Technical and Artistic Aspects of Videotape in Psychiatric Teaching." J. Nerv. Ment. Dis. 144:207–223.
141. ———, "The Use of Television Videotape in a Therapeutic Community for Adolescents Involved with Drugs." Paper presented at the Western Divisional Meeting of the American Psychiatric Association, Los Angeles, California, October 1967.
142. ———, "Innovative Uses of Videotape on a Psychiatric Ward." Hosp. and Comm. Psychiat. 19:129–133.
143. ———, "Television as Participant Recorder." Amer. J. Psychiat. 124:43–49.
144. ———, "The Undisguised Camera in Psychiatry." Visual/Sonic Medicine 3:5–11.
145. ———, "Use of Television Monologues with Adolescent Patients." J. Amer. Psychiat. 126:1760–1766.
146. ———, "A New Tool in Psychiatry: Videotape Confrontation." Roche Medical Image and Commentary 11:20–23.
147. ———, "TV Soliloquies Help Young Drug Users." Today's Health, April 1970, pp. 64–65.
148. ———, "Use of the Television Monologue with Adolescent Psychiatric Patients." Amer. J. Psychiat. 126:1760–1766.
149. ———, "Feedback: TV Monologue Psychotherapy." Radical

Software, summer issue, no. 4, p. 11, 1971.
150. ———, and Rynearson, R. R. Videotherapy. Addictions 20:40–46.
151. Woody, R. H. et al. "Stimulated Recall in Psychotherapy Using Television and Videotape." Amer. J. Clin. Hypn. 7:234–241.
152. ——, "Teaching Hypnosis with Video Tapes." Amer. J. Clin. Hypnosis 8:111–113.
153. ———, "Clinical Suggestion and the Videotaped Vicarious Desensitization Method." Amer. J. Clin. Hypnosis 11:239–244.
154. ———, and Schauble, P. G. "Desensitization of Fear by Video Tapes." J. Clin. Psychol. 25:102–103.
155. ———, and ——— "Videotaped Vicarious Desensitization Methodology and Research." J. Nerv. Ment. Dis. 148:281–186.
156. ———, "Clinical Suggestion in Videotaped Psychotherapy: A Research Progress Report." Am. J. Clinical Hypnosis 14:32–37.
157. Younge, K. A. "The Use of Closed Circuit Television for the Teaching of Psychotherapeutic Interviewing to Medical Students." Can. Med. Ass. J. 92:747–751.

INDEX

A

Acceleration of psychotherapy, 200
Active-active mode, 157
Acitive-passive-active mode, 157
Active-passive mode, 157
Active-passive-passive mode, 157
Acupuncture anesthesia, 11,1, 112
Adler, 156
Adolescents, 209
Affleck, D. C., 200
Africa, 110, 116
Alcoholic, 68, 206
Alcoholism, 202
Alger, I., 200, 203
Alktre, A. S., 200
Allergist, 115
AMA, 83
Ambigram, 160
Ambigraph, 160
Ambistory, 160
Ambivalence, 160
American Academy of Family Practice, 124
American Academy of General Practice, 124
American Medical Association, 114
Amnesia, 160
Analysand, 160
And-rule, 161
Anderson, C. J., 200
Anderson, J., 200
Anniversary phenomenon, 83
Antabuse, 36
APA, 63
Apostoles, E. F., 200
Artistic aspects of videotape, 209
Asia, 110, 116
Assagioli, 51, 52, 156, 171, 172
Audio dubbing, 148

B

Baby-sitting, 114
Bahnson, C., 200
Bailey, K. G., 200
Baird, John, 127
Barenboum, C., 201
Barlow, D. J., 200
Barnes, L. H., 200
Baugh, J. R., 205
Behavior therapy, 46
Benschoter, R. A., 200
Berger, B., 207
Berger, M. M., 201
Bergman, A. C., 201
Bertalanffy, Ludwig von, 131
Binary code, 161
Bit, 161
Blancheri, R. L., 201
Blount, H. P., 201
Bodin, A. M., 201
Boolean algebra, 161
Booster session, 24
Booster telefusion, 53, 54
Booster telefusion technique, 108
Brain, 60
Brain-mind, 97
Brayton, D., 201
Brown, R., 101, 102
Bruhn, J. G., 200
Brunse, A. J., 200

C

Camera cable, 146
Cardiologist, 115
Cartesian dictum, 95
Cassette, 144
Catalyzer, 161
CCTV, 143

CCTV cameras, 145
CCTV monitors, 149
Cecil, L., 200
Cerebral cortex, 98, 99
Cerletti, 51, 92
Chandler, M. J., 201
Channel, 161
Chemolytic, 161
Chemotherapy, 42, 47
Chernell, E., 205
Chief complaint, 3, 34, 37, 42
Christopher, D., 206
Chronic patients, 202
Circulating memory, 94, 111, 161
Clear, 162
Client satisfaction, 202
Closed circuit TV, 203
Code, 162
Coffee shop, 59, 61, 122, 162
Cohen, M. J., 204
Community mental health, 201
Confrontation, 209
Conjoint marital therapy, 200
Consciousness, 83, 95, 162
Consciousness and personality, 99
Continuity, 85
Contradictions-in-identity, 162
Controlling personality, 162
Cook, O., 204
Coone, J. G., 202
Cornelison, F. S., 202
Cosman, 162
Cost, 80
Cottreel, T. B., 205
Couch, 59
Couch analysis, 12, 55
Couch therapy, 12, 157
Cozean, C. E., 202
Crisis, 122
Cybernetics, 163
Cybernetic system, 7
Cybernetic telefusion, 8
Cycloptic TV monitor, 5
Czajkoski, E. H., 202

D

Daedalus, 65
Danet, B. N., 202
Darwinian, 101
Davies, T. S., 202
Davis, C. R., 205
Davison, W. T., 206
Delaware State Hospital, 202
Delinquent adolescents, 204
Democritus, 172
Depression, 41
Depression Inventory Measure, 190
Depressive neurosis, 41
Desensitization, 204, 210
Diabetes, 77
Diabetics, 77
Dialectics, 51
Diethem, O., 171
Differential diagnosis, 33, 36, 39, 41
D.I.M., 190
Discipline children, 120
DNA, 94
Doctor-patient relationship, 79
Dominant personality, 163
Drug therapy, 13, 51
Dwyer, T. E., 202
Dynamic thresholds, 163

E

Eaton, M. T., 201, 206
Eclectic, 97, 129, 130
Eclectic psychiatrist, 60
Ego, 163
EIAJ 147, 148
EIAJ color, 144
Einstein, 47, 97
Einsteinian, 47, 99
Electroencephalogram, 33, 36, 38, 41, 43
Electrolytic therapy, 50, 100
Electrotherapy, 13, 47
Elighman, A., 202
ELT, 163
Emmitt, T., 208
Emotional stability, 11
Energy, 51, 60, 100, 163
Energy of transformation, 163
Erase, 164
Erasing, 164
Erasure, 164
Erb, 92
Error, 164
Examination, 1, 3
Existential neurosis, 173
Eysensck, 46

F

Families for Psychosynthesis, Inc., 61

Family diagnosis, 205
Family history, 4, 25, 32, 35, 37, 40, 43, 183
Family medicine, 73, 117
Family physician, 113
Family practice, 113, 114
Family psychiatrist, 113
Family psychiatry, 7, 60, 208
Family therapists, 201
Family therapy, 13, 39, 50, 59
Feedback, 47, 105, 164, 205
Feristein, C., 202
FM-radio, 204
Forensic, 156
Forman, L. D., 206
Forrest, D. V., 202
Free associate, 55
Free association, 158, 164
Free communication, 54, 158, 164
Freed, H., 203
Frequency response, 148
Freud, 51, 63, 65, 70, 92, 101, 102, 127, 156, 157
Freudian, 104
Furman, S., 202

G

Gabor, D., 98, 101, 102
Gallant, D. M., 201
Galvin, R., 202
Gant, H. M., 202
Gauson, F. F., 202
Geertsma, R. H., 203, 206
Gelso, C. J., 202
General practitioner, 113
Gentry, K. D., 206
Gibson, R., 127
Gildfield, M. D., 203
Gioscia, V., 203
Gladfelter, J., 203
Goal-directed therapy 164
Goldfield, M. D. 203
Gould, J., 203, 207
Great Britain, 110
Greeks, 65
Green, J. H., 203
Grienspan, S., 201
Griffin, G. J., 203
Grinker, Roy, 131
Group and family videotherapy, 42
Group psychotherapy, 202, 207
Group therapy, 13, 36, 39, 47, 50, 63, 73, 84, 117, 200, 202
Grueneberg, P. B., 203
Guerilla television, 5
Gundel, S., 206

H

Hanel, L. C., 205
Happiness, 12
Harris, J. J., 203
Harrison, C. A., 205
Hatton, R. L., 7

Heads, 148
Hebephrenic type, 33
Henry, G. W., 171
Hewitt, C., 204
Hirsh, H., 203
Historical purview, 203
History, 1
Hogan, P., 200, 203
Holmes, D. J., 203
Hologram, 8, 95, 111
Hologram model, 98, 208
Horizontal resolution, 146, 148
Hospitalization, 26, 39
Hoye, C. L., 200
Hubbard, O. E., 205
Human cortex, 105
Hyperactive children, 202
Hypnosis, 65
Hypnosis with video, 210
Hypnotherapy, 10

I

Impedance, 72, 164
Information, 164
Ingham, C. G., 200
Initial history, 3
Initial history and examination, 10
Initial interpretation, 33, 39
Input impedance, 148
Input-stayput-output system, 143
Institute of Psychoanalysis, 171
Intensive supervision of psychotherapy, 203

J

Jacobs, A., 206
Japan, 128

Jaramillo, James D. C., 203
Jastrow, 66
Jastrow-boring phenomenon, 165
Journey of the Stone Monkey, 69, 173
Journey to the West, 173
Jung, C. G., 171
Juvenile delinquents, 205

K

Kagan, N., 203, 206
Kinetic memory, 94, 165
Kinsinger, R. E., 203
Kolb, L. C., 204
Kornfled, D. S., 204
Krathwohl, D. R., 203
Krystal, H., 204
Kubie, L. S., 204
Kurland, A. A., 204

L

Lansing, 79
Lautch, J., 204
Law of Universal Truth, 66, 165, 173
Laybourne, P. D., 206
Learning, 10
Lens attached, 146
Levey, R., 203
Lind, J., 204
Liston, E. H., 203
Loeb, G., 205
Long-term psychiatric patients, 200
Longuet-Higgins, 101
Loosening, 165
Lucretius, 172
Lytic, 165
Lyzer, 165

M

Maclay, D. T., 204
Mann, J., 204
Manpower problem, 115
Manual control type, 146
Marin, G. L., 204
Marital therapy, 36, 42, 45
Marvit, R. C., 204
Mass communication, 6
Masserman, Jules, 208
Master program, 165
Matrix, 165
McClaughin, D. G., 204
McCulloch, W., 94
McDermott, J. F., 204
McGuire, F. L., 204
McLuhan, M., 105, 112, 130
McNiff, A., 204
Meadow, R., 204
Medical economics, 132
Medical history, 31
Medical manpower, 81
Medical students, 210
Medical television, 208
Medical TV consultants, 144
Medico-legal, 156
Melnick, J., 204
"Memini ergo sum," 51
Memory, 94, 165
Mental health community, 61
Mental health mini-community, 50
Merrill, L. R., 201
Merrital, H. H., 202
Michaux, M. H., 204
Michigan Institute of Psychosynthesis, 23, 62, 113
Michigan Radiologic Society, 134
Michigan relative value reports, 132
Microphone, 150, 155
Miller, D., 205
Miller, R., 203
Millerson, G., 205
Mind, 60
Mind fusion, 166
Minicommunity, 101
MMPI, 41
Model, 100
Modified medical schedule (Anxiety scale), 192
Monitor, 149
Moore, F. J., 205
Moreno, 63, 64, 66, 67, 68, 205
Morita, Akio, 128
Most-similarity, 166
Multiple image, 201
Multipurpose television, 201

N

Name, 166
Nash, L. R., 208
Neal, C. D., 207
Neoname, 166
Neurogenic interference theory, 112

Neurological examination, 32, 35, 38, 41, 43, 188
Newborn, 100
Nursing instruction, 203

O

Obstetrician, 115
Oedipus Complex, 51, 94
O'Hara, H., 205
Omnidirectional microphone, 149
Onder, J. J., 205
Operating ambient temperature, 146
Operating humidity, 146
Ordinary session, 24
Ordinary telefusion, 53, 54, 108
Orwell, George, 5
Output impedance, 148
Over, H. R., 204
Overall, J. E., 205

P

Paralegal, 23
Paramedical, 23
Parapsychiatric, 23
Parents, 121
Participant recorder, 209
Pascal, G. R., 205
Past disturbance, 34, 37
Past problems, 3, 40, 42
Patient contact form, 177
Patient-therapist alliance, 166
Pattern recognition, 52, 166
Pattern recognition by identify theory, 166
Pavlov, I. P., 51, 103, 172
Peace of mind, 12
Pederson, A. M., 201
Pediatrician, 115
People's Republic of China, 110
PERF Model, 1, 9, 22, 104, 105, 168, 169
Perlmitten, M., 205
Person, 167
Personal history, 3, 16, 19, 32, 34, 37, 40, 43, 182
Personality, 87, 100, 167
Personality development, 13
Personality disorder, 36, 39
Personality transformation, 97, 167
Phenomenon of retrograde amnesia, 93
Physical examination, 189
Pitman, A. J., 203
Playback, 200
Pledge of psychosynthesis, 167
Pneumonia, 72
POPF Model, 1, 9, 22, 24, 105, 106, 107, 108, 167
Postlytic, 170
Postlytic family therapy, 50
Power consumption, 146, 148
Power source, 146, 148, 149
Prelytic, 170
Prename, 170
Present disturbance, 34, 37, 42
Previous treatment, 3, 18, 34, 37, 40
Primary telefusion, 53, 108
Primary transnomation, 170
Principle of most-similarity, 95
Principle of psychosynthesis, 170
Private practice, 201
Problem questionnaire, 196
Procedure of immediate reprogramming, 94
Programmable, 170
Programmer, 109, 170
Pro-personality, 170
Psychiatric examination, 32, 35, 38, 40, 43, 186
Psychiatric hospital, 19, 209
Psychiatric prognostic scale, 194
Psychiatrist, 130
Psychiatry, 129
Psychoanalysis, 46, 129, 157, 204
Psychoanalyst, 46, 127, 130, 131
Psychoanalytic interviews, 206
Psychocybernetic model, 104
Psychodrama, 64, 203, 205
Psychodramatist, 64
Psychological tests, 33, 36, 38, 41
Psychology, 51
Psychology course, 202
Psycholysis, 170
Psychomutation, 100
Psychopathology, 204
Psychoschists, 170
Psychosomatic medicine, 124, 208
Psychosphere, 98, 111
Psychosynthesis, 46, 51, 60, 92, 129, 157, 171, 208
Psychosynthesist, 97, 128, 130, 131, 172
Psychosynthetic, 172
Psychotherapy, 12, 47, 133

R

Radiology, 30, 117
Ramey, J. W., 205
Random-dot hologram, 98
Raulings, F. T., 202
Rawis, J. R., 205
Reactivation, 172
Recording time, 148
Reivich, R. S., 202, 203, 206
Repressed material, 201
Required illumination, 146
Research paradigm, 204
Reset, 172
Resistance, 72
Resnik, H. L., 206
Resnikoff, A., 206
Resolution, 145, 149
Retrograde amnesia, 93
RF adaptor, 147
Richter, H. E., 206
RNA, 94
Robinson, M., 206
Rocky Mountain Bioengineering Symposium, 8
Rogers, A. H., 206
Rogus, A., 206
Role-plays, 200
Romano, M. T., 208
Rose, J., 7
Ruhe, D. S., 206
Russell, Bertrand, 67
Russell, S. C., 205
Ryan, J. H., 202, 206
Rynearson, R. R., 206, 210

S

Sainanto, H. K., 200
Sauces, A., 8
Scanning standard, 146
Schaefer, H. H., 206
Schauble, P. G., 206, 210
Schiff, S. B., 206
Schisis, 172
Schizo-affective schizophrenia, 44
Schizoid personality, 44
Schizophrenic, 33
Schuyler, D., 206
Scintillating memory, 94
Sclare, A. B., 206
Scott, M., 207
Second personality, 172
Self-concept, 201
Self-controntation, 200, 201, 202, 206
Self-instructional mode, 205
Self-medication, 19
Self-observation, 203
Self-visualization, 80
Sensory-motor input, 111
Setback, 172
Sethna, E. R., 207
Shaeffer, S., 207
Shakespeare, 69
Shannon, 51, 101, 103
Sherman, B., 201
Shostrom, E. L., 207
Signal-to-noise ratio, 146, 148
Skamia, H. J., 207
Smith, K. U., 203, 207
Smith, P., 201
Smith, T. J., 207
Sobell, L., 206
Sobell, M., 206
Social psychiatry, 47
Sociosynthesis, 172
Socrates, 67
Solid state devices, 146
Solow, C., 207
Sony, 128
South America, 110, 116
Soviet Union, 110
Sower, W. T., 200
Spalding, J., 201
Special effects generator, 146
Static memory, 173
Steady service, 60
Stereogram, 105
Stereotelevision, 203
Stigall, T. T., 204
Stimulated recall, 203
Stigall, T. T., 204
Stimulated recall, 203
Stoller, F. H., 207
Stone Monkey Syndrome, 173
Stroms, M. D., 207
Struggle, 173
Stuart, B. R., 205
Students, 205
Sub-personality, 173
Suess, J. F., 207
Suicide, 206
Supervision of group psychotherapists, 203
Sync, 146
Synthesand, 53

Synthesant, 173
Synthesis, 51, 69
Synthesist, 66, 173
System function, 174
Systemic review, 4, 26, 32, 35, 37, 40, 43
Systems theory, 207
Szasz, 46

T

Tamerin, J., 202
Tape length, 148
Tape speed, 148
Tape width, 148
Tausig, T. M., 202, 207
Teaching by video-tape, 206
Teaching medical students, 203
Teaching psychiatry, 203, 204
Technique of memory reactivation, 93
Telefusion, 46, 51, 143, 157, 174
Telefusion techniques, 108
Telepsychiatry, 202
Telesynthesis, 174
Televised audiovisual reproduction, 204
Television, 60, 71, 156
Television and society, 207
Television hypnosis, 207
Television production, 205
Tentative prognosis, 33, 36, 39
Therapeutic films, 208
Thermonuclear bomb, 51
Thompson, J. W., 208
Threshold, 174
Tien, H. C., 7, 103, 112, 208
Time-series, 99
Tims, A. R., 204
Toeman, Z., 208
Torkelson, L. O., 208
TOTE Model, 174
Transformation, 99
Transformation of personality, 96
Transnomation, 50, 72, 175
Trend and content questionnaire, 4, 28
Trethowan, W. H., 208
Trull, S. G., 208
TV booster session, 7, 174
TV-couch session, 174
TV-cybernetic, 93, 112
TV-cybernetic system, 101, 208
TV group booster, 84
TV in the teaching of psychiatry, 204
TV in psychiatric nursing, 202
TV-linked session, 7, 10
TV-monitored approach, 6
TV-monitored evaluation forms, 2, 178
TV-monitored form, 14
TV-monitored interviews, 16
TV-monitored medical history, 2, 9, 16, 24
TV monologue psychotherapy, 210
TV programmer, 6, 22
TV sessions, 30
TV soliloquies, 209
TV-tape, 174
TV-taped couch session, 7
Two-person interview, 205
Two-personality theory, 175
Two-way TV, 100

U

Undisguised camera, 209
Unidirectional dynamic microphone, 149
United States, 110
Universal truth, the law of, 175

V

Vicarious desensitization, 210
Video art therapy, 204
Video asa medium, 206
Video camera, 143
Video cartridge, 151, 153
Video cartridge tape recorder, 147
Video cassette editing system, 154
Video cassette player, 151
Video cassette psychiatry, 202
Video feedback, 202
Video frequency response, 146
Video input, 149
Video-oriented psychiatric practice, 203
Video output, 146
Video recorder, 143
Videocorder, 144
Videoist, 23, 175
Videologist, 23, 117, 175
Videology, 6, 23, 109, 156, 158, 176, 208
Videology economics, 132

Videosynthesis, 102, 104, 176
Videosynthesist, 61
Videosystem, 9, 143
Videotape, 83, 176, 204
Videotape confrontation, 206
Videotape player, 151
Videotape recorder, 153
Videotape stimulation, 204, 205
Videotape techniques, 201
Videotechniche, 51, 143
Videotherapy, 102, 127, 176
Vidicon, 145
Vidicon camera, 145
Vidicon tube, 146
Biewfinder, 146
VTR, 143
VTR microphones, 149
VTR record, 145

W

Warren, F. Z., 208
Watts, J. C., 209
Wayne, G. J., 203
Weight control, 74
Weiss, R., 207
West, M. J., 205
Westlake, R., 201
White, L. A., 52
Wiener, W., 51, 68, 92, 97, 101, 110, 172
Wilhite, J. A., 209
Wilmer, H. A., 206, 209
Wittson, C. L., 200
Wolpe, 46
Woody, R. H., 210
Woolf, V., 171
World Congress of Psychiatry, 124
World Health System, 110
Wu Chengen (WWUD Chengsj Enxn), 173
Wulfsohn, N. L., 8

X

X-ray, 117, 156
X-ray machine, 109

Y

Younge, K. A., 210

Z

Zeus, 67
Zweig, S., 171

PHOTO INDEX

Author, the Videologist 1
TV-Monitored Medical Forms 9
Primary Telefusion Couch Therapy 53
Patient in Primary Telefusion Couch Therapy 55
Telefusion Couch Feedback 56
First TV-linked Group Therapy 63
Second TV-linked Group Therapy 73
Third TV-linked Group Therapy 84
The POPF Model 106–107
TV-linked Group Therapy Revisited 117
The PERF Model 168–169